Solutions Manual to Accompany

Principles of Highway Engineering and Traffic Analysis

Second Edition

Fred L. Mannering
University of Washington

Walter P. Kilareski
The Pennsylvania State University

John Wiley & Sons, Inc.
New York / Chichester / Weinheim / Brisbane / Singapore / Toronto

SOLUTIONS: CHAPTER 2

PROBLEM 2.1

$$R_a = \frac{\rho}{2}C_D A_f V^2$$

$$= \frac{1.2256}{2}(0.29)(1.9)(160 \times 0.2778)^2$$

$$= 667.08 \text{ N}$$

$$\text{Power} = R_a V = 667.8(160 \times 0.2778) = 29682.37 \text{ Nm}$$

PROBLEM 2.2

at V_m power $= R_a V_m + R_{rl} V_m$

additional power $= W \times 1000/9$

$$11.11W = (1.2256/2)(0.30)(2)(160 \times 0.2778)^3$$

$$+ 0.01\left(1 + \frac{(160 \times 0.2778)}{44.73}\right)(9300 + W)(160 \times 0.2778)$$

$$11.11W = 32286.92 + 8241.27 + 0.886 W$$

$$= 367.69 \text{ N}$$

Total vehicle weight $= 9667.69 \text{ N}$

PROBLEM 2.3

$$\text{rear } F_{max} = \text{ front } F_{max}$$

$$\frac{\mu W\left(l_f - f_{rl}h\right)/L}{1 - \mu h/L} = \frac{\mu W\left(l_f + f_{rl}h\right)/L}{1 + \mu h/L}$$

$$\frac{0.6(11000)\left(l_f - 0.01(55)\right)/305}{1 - 0.6(55)/305} = \frac{0.6(11000)\left(305 - l_f + 0.01(55)\right)/305}{1 + 0.6(55)/305}$$

$$43.7913 l_f = 5979.6956$$

$$l_f = 136.55 \text{ cm}$$

PROBLEM 2.4

$$F = ma + R_{rl} = \text{rear } F_{max} = \frac{\mu W (l_f - f_{rl} h)/L}{1 - \mu h/L}$$

$$\frac{13300}{9.807}(4.6) + 0.01(13300) = \frac{\mu(13300)(355 + 0.01(50))/510}{1 - \mu(50/510)}$$

$$6375.9 = \frac{9244.8\mu}{1 - 0.01\mu}$$

$$\mu = 0.645$$

PROBLEM 2.5

$$F = ma + R_{rl} = \frac{\mu W (l_f - f_{rl} h)/L}{1 - \mu h/L}$$

$$\frac{13300}{9.807}(11.77) + 0.01(13300) = \frac{0.95(13300)(l_f - 0.01(178))/510}{1 + 0.95(178/510)}$$

$$16093 = \frac{24.77 l_f - 44.1}{0.6684}$$

$$l_f = 436.04 \text{ cm}$$

PROBLEM 2.6

$$= \frac{2\pi r n_e (1 - i)}{\varepsilon_0}$$

$$= \frac{(6.28)(0.38)(50)(1 - 0.035)}{3.5} = 32.9 \text{ m/s}$$

$$R_a = \frac{\rho}{2} C_D A_f V^2 = \frac{1.2256}{2}(0.35)(2.0)(32.9)^2 = 465.00 \text{ N}$$

$$R_{rl} = f_{rl}W = 0.01\left(1 + \frac{32.9}{44.73}\right)(13300) = 230.81 \text{ N}$$

$$\gamma_m = 1.04 + 0.0025(3.5)^2 = 1.071$$

$$F_e = \frac{M_e \varepsilon_0 n_d}{r} = \frac{(340)(3.5)(0.9)}{0.38} = 2818.4 \text{ N}$$

$$F_{net} = F - \sum R = \gamma_m ma$$

$$\text{so, } a = \frac{2818.4 - 465.00 - 230.81}{1.071(13300/9.807)} = 1.46 \text{ m/s}^2$$

PROBLEM 2.7

$$= \frac{2\pi r n_e (1 - i)}{\varepsilon_0}$$

$$= \frac{(6.28)(0.355)(3500/60)(1 - 0.035)}{3.2} = 39.22 \text{ m/s}$$

$$R_a = \frac{\rho}{2} C_D A_f V^2 = \frac{1.2256}{2}(0.35)(2.3)(39.22)^2 = 758.80 \text{ N}$$

$$R_{rl} = f_{rl}W = 0.01\left(1 + \frac{39.22}{44.73}\right)(11000) = 206.45 \text{ N}$$

$$F_e = \frac{M_e \varepsilon_0 n_d}{r} = \frac{(270)(3.2)(0.9)}{0.355} = 2190.4 \text{ N}$$

$$F = W\sin\theta + R_{rl} + R_a$$

$$2190.4 = 11000\sin\theta + 758.80 + 206.45$$

$$\sin\theta = 0.11138 \approx G = 11.14\%$$

PROBLEM 2.8

$$F_e = \frac{M_e \varepsilon_0 n_d}{r}$$

$$= \frac{(270)(3)(0.9)}{0.355}$$

$$= 2053.5 \text{ N}$$

$$f_{rl}W = 0.01\left(1 + \frac{242 \times 0.2778}{44.73}\right)(11000) = 275.33 \text{ N}$$

$$R_a = \frac{\rho}{2}C_D A_f V^2 = \frac{1.2256}{2}(C_D)(2.3)(242 \times 0.2778)^2 = 6370.03 C_D$$

$$F_e = f_{rl}W + R_a$$

$$2053.5 = 275.33 + 6370.03 C_D$$

$$C_D = 0.279$$

PROBLEM 2.9

$$F_{max} = \frac{\mu W(l_f - f_{rl}h)/L}{1 - \mu h/L} = \frac{0.75(11000)(76 - 0.01(50))/203}{1 - 0.75(50)/203} = 3763.6 \text{ N}$$

For maximum torque

$$\frac{dM_e}{dn_e} = 6 - 0.09 n_e = 0$$

$$n_e = 66.67$$

$$M_e = 6(66.67) - 0.045(66.67)^2 = 200 \text{ Nm}$$

$$F_e = \frac{M_e \varepsilon_0 n_d}{r} = \frac{(200)(11)(0.75)}{0.355} = 4647.89 \text{ N}$$

$$R_{rl} = f_{rl}W = 0.01(13300) = 133 \text{ N}$$

$$\gamma_m = 1.04 + 0.0025(11)^2 = 1.34$$

$$a = \frac{F_{max} - f_{rl}W}{\gamma_m m} = \frac{3763.6 - 133}{1.34(11000/9.807)} = 2.431 \text{ m/s}^2$$

PROBLEM 2.10

$$\text{power} = 2\pi M_e n_e = 6.28(6n_e^2 - 0.045 n_e^3) = 37.68 n_e^2 - 0.2826 n_e^3$$

For maximum power

$$\frac{dP}{dn_e} = 75.36 n_e - 0.8478 n_e^2 = 0$$

$$n_e = 88.89$$

$$= \frac{2\pi r n_e (1 - i)}{\varepsilon_0} = \frac{6.28(88.89)(0.355)(0.965)}{2} = 95.62 \text{ m/s or } 344.23 \text{ km/h}$$

PROBLEM 2.11

R_a, R_{rl}, and γ_m are as before

$$F_e = \frac{M_e \varepsilon_0 n_d}{r} = \frac{(130)(4.5)(0.8)}{0.33} = 1418.18 \text{ N}$$

$$a_{max} = \frac{F - \sum R}{\gamma_m \text{m}} = \frac{1418.18 - 152.48}{1.091(13300/9.807)} = 0.855 \text{ m/s}^2$$

Rear – wheel drive

$$F_{max} = \frac{\mu W (l_f - f_{rl} h)/L}{1 - \mu h/L} = \frac{0.2(13300)(152.5 - 0.011(50))/305}{1 - 0.2(50)/305} = 1370.13 \text{ N}$$

$$a_{max} = \frac{1370.13 - 152.48}{1.091(13300/9.807)} = 0.823 \text{ m/s}^2 < 0.855 \text{ m/s}^2$$

Front – wheel drive

$$F_{max} = \frac{\mu W (l_1 + f_{rl} h)/L}{1 + \mu h/L} = \frac{0.2(13300)(152.5 + 0.011(50))/305}{1 + 0.2(50)/305} = 1292.4 \text{ N}$$

$$a_{max} = \frac{1292.4 - 152.48}{1.091(13300/9.807)} = 0.77 \text{ m/s}^2 < 0.855 \text{ m/s}^2$$

PROBLEM 2.12

$$F_e = \frac{M_e \varepsilon_0 n_d}{r} = \frac{(M_e)(10)(0.8)}{0.355} = 22.53 M_e$$

$$F_{max} = \frac{\mu W (l_f - f_{rl} h)/L}{1 - \mu h/L} = \frac{\left[0.8(8900 + 13 M_e)((140 - 0.36 M_e) - 0.01(55))\right]/255}{1 - 0.8(55)/255} = 1370.13 \text{ N}$$

$$= (255)(22.53 M_e)(0.827) = (7120 + 10.4 M_e)(139.45 - 0.36 M_e)$$

$$3.74 M_e^2 + 3638.32 M_e - 992884 = 0$$

$$M_e = 222.16 \text{ Nm}$$

and engine weight $= 222.16(3) = 2888.1 \text{ N}$

5

PROBLEM 2.13

$$K_a = \frac{1.2256}{2}(0.45)(2.3) = 0.6342$$

$$S = \frac{1.04(11000)}{2(9.807)(0.6342)}\ln\left[1 + \frac{0.6342(145 \times 0.2778)^2}{(1.0)(0.7)(11000) + (0.0145)(11000) - 11000\sin(5.71°)}\right]$$

$$= 130.23 \text{ m}$$

so the difference is $139.89 - 130.32 = 9.66\text{m}$

PROBLEM 2.14

$$K_a = \frac{1.2256}{2}(0.4)(2.4) = 0.588$$

a. considered

$$76 = \frac{1.04(15600)}{2(9.807)(0.588)}\ln\left[1 + \frac{0.588V^2}{(0.5)(0.78)(15600) + (0.015)(15600)}\right]$$

$$1.0555 = \left[1 + \frac{0.588V^2}{6318}\right]$$

$$V = 24.42 \text{ m/s or } 87.92 \text{ km/h}$$

b. ignored

$$76 = \frac{1.04V^2}{2(9.807)(0.78(0.5) + 0.015)}$$

$$V = 24.09 \text{ m/s or } 86.74 \text{ km/h}$$

PROBLEM 2.15

$$S = \frac{\gamma_b V_1^2}{2g(\eta_b\mu + f_{rl} \pm \sin\theta_g)} = 61 = \frac{1.04(100 \times 0.2778)^2}{2(9.807)(\eta_b(0.75) + 0.018)}$$

$$897.34\eta_b + 21.54 = 802.60$$

$$\eta_b = 0.8704 \text{ or } 87.04\%$$

PROBLEM 2.16

$$f_{rl} = 0.01\left(1 + \frac{60 \times 0.2778}{44.73}\right) = 0.137$$

$$S = \frac{\gamma_b\left(V_1^2 - V_2^2\right)}{2g\left(\eta_b\mu + f_{rl} \pm \sin\theta_g\right)} = 90 = \frac{1.04(120 \times 0.2778)^2}{2(9.807)\left(\left(0.80 - \frac{W}{44500}\right)(0.95) + 0.0137 - 0.04\right)}$$

$$90 = \frac{1155.74}{14.91 - 0.00042W - 0.5158}$$

$$W = 3696.77 \text{ N}$$

PROBLEM 2.17

$$S = \frac{\gamma_b W}{2gK_a}\ln\left[\frac{\eta_b\mu W + K_a V_1^2 + f_{rl}W \pm W\sin\theta_g}{\eta_b\mu W + K_a V_2^2 + f_{rl}W \pm W\sin\theta_g}\right]$$

$$f_{rl} = 0.018$$

$$\eta_b\mu W = 0.8(0.85)(15600) = 10608$$

$$K_a = (1.2256/2)(0.5)(2.3) = 0.70472$$

on level surface

$$45 = \frac{1.04(15600)}{2(9.807)(0.70472)}\ln\left[\frac{10608 + 0.70472(130 \times 0.2778)_1^2 + 0.018(15600)}{10608 + 0.70472V_2^2 + 0.018(15600)}\right]$$

$$0.03834 = \ln\left[\frac{12726.45}{10888.8 + 0.70472V_2^2}\right]$$

$$.0391(10888.8 + 0.70472V_2^2) = 11807.91$$

$$_2 = 25.96 \text{ m/s or } 93.45 \text{ km/h}$$

on 5% grade

$$45 = \frac{1.04(15600)}{2(9.807)(0.70472)}\ln\left[\frac{10608 + 0.70472(36.114)_1^2 + 0.018(15600) + 0.05(15600)}{10608 + 0.70472V_2^2 + 0.018(15600) + 0.05(15600)}\right]$$

$$0.03834 = \ln\left[\frac{13506.45}{11668.8 + 0.70472V_2^2}\right]$$

$$1.0391(11668.8 + 0.70472V_2^2) = 12587.153$$

$$V_2 = 25.14 \text{ m/s or } 90.5 \text{ km/h}$$

PROBLEM 2.18

$f = 0.30$

$$d = \frac{V_1^2 - V_2^2}{2g(f \pm G)} = \frac{(90 \times 0.2778)^2 - (56 \times 0.2778)^2}{2(9.807)(0.30)} = 65.104 \text{ m}$$

$180 - 65.104 = 114.896$ m for reaction distance

$d_p = V_1 t_p = (90 \times 0.2778)t_p$

$t_p = 4.595$ s

PROBLEM 2.19

at 100 km/h

$$\frac{\gamma_b V_1^2}{2g} = \frac{1.04(100 \times 0.2778)^2}{2(9.807)} = 42.49$$

$$f_{rl} = 0.01\left(1 + \frac{50 \times 0.2778}{44.73}\right) = 0.0131$$

$$00(2.5)(0.2778) + \frac{42.49}{\eta_{b1}\mu + 0.0131} = 100(2.0)(0.2778) + \frac{42.49}{\eta_{b2}\mu + 0.0131}$$

$$3.89 = \frac{42.49}{\eta_{b1}0.8 + 0.0131} + \frac{42.49}{\eta_{b2}0.8 + 0.0131}$$

with $\eta_{b2} = 0.75$

$$55.41 = \frac{42.49}{\eta_{b1}0.8 + 0.0131}$$

$\eta_{b1} = 0.94$ or 94%

PROBLEM 2.20

$$d = \frac{V_1^2}{2gf} = \frac{(110 \times 0.2778)^2}{2(9.807)(0.28)} = 170.03 \text{ m}$$

$180 - 170.03 = 9.97$ m for reaction time

$d_p = V_1 t_p = 9.97 = (110 \times 0.2778)t_p$

$t_p = 0.326$ s, therefore student is probably incorrect

PROBLEM 2.21

$$d_p = V_1 t_p = (90 \times 0.2778)(2.5) = 62.51 \text{ m}$$

$$d = 140 - 62.51 = 77.49 = \frac{V_1^2}{2gf} = \frac{(90 \times 0.2778)^2}{2(9.807)(0.30 + G)}$$

$$1519.89G + 455.97 = 625.1$$

$$G = 0.1113 \text{ or } 11.13\%$$

PROBLEM 2.22

while sober,

$$d_s = V_1 t_p + \frac{V_1^2}{2gf} = (90 \times 0.2778)(t_p) + \frac{(90 \times 0.2778)^2}{2(9.807)(0.30)}$$

$$t_p = 2.15 \text{ s}$$

after drinking

$$160 = (90 \times 0.2778)(t_p) + \frac{(90 \times 0.2778)^2 - (55 \times 0.2778)^2}{2(9.807)(0.30)}$$

$$t_p = 3.74 \text{ s}$$

SOLUTIONS: CHAPTER 3

PROBLEM 3.1

$$a = \frac{G_2 - G_1}{2L} = \frac{0.065 + 0.035}{2(350)} = 0.000143$$

$$b = -0.035$$

at low point

$$\frac{dy}{dx} = 2ax + b = 0.000286x - 0.035 = 0$$

$$x = 122.5 \text{ m}$$

for elevation of low point

$$y = ax^2 + bx + c$$

$$y = 0.000143(122.5)^2 - 0.035(122.5) + 280$$

$$y = 277.858 \text{ m}$$

PVI sta $3+040.000 + 0+175 = 3+215.000$

elev $280 - 0.035(175) = 273.875$ m

PVT sta $3+040.000 + 0+350 = 3+390.000$

elev $273.875 + 0.065(175) = 285.250$ m

low point sta $3+040.000 + 0+122.5 = 3+162.500$

elev 277.858 m

PROBLEM 3.2

$$a = \frac{G_2 - G_1}{2L} = \frac{-0.025 - 0.04}{2(160)} = -0.000203$$

$$b = 0.04$$

at high point

$$\frac{dy}{dx} = 2ax + b = -0.000406x + 0.04 = 0$$

$$x = 98.522 \text{ m}$$

for elevation of high point

$$c = 411 - 0.04(80) = 407.8 \text{ m}$$

$$y = ax^2 + bx + c$$

$$y = -0.000203(98.522)^2 + 0.04(98.522) + 407.8$$

$$y = 409.771 \text{ m}$$

PVC sta $4+310.000 - 0+080 = 4+230.000$

elev 407.8 m

PVT sta $4+310.000 + 0+080 = 4+390.000$

elev $411 - 0.025(80) = 409$ m

high point sta $4+230.000 + 0+098.522 = 4+328.522$

elev 409.771 m

PROBLEM 3.3

$$a = \frac{G_2 - G_1}{2L} = \frac{-0.0108 - 0.012}{2(150)} = -0.000076$$

$b = 0.012$

PVC sta $11+000 - 0+075 = 10+925.000$

elev $1095.2 - 0.012(75) = 1094.3$ m

at station $11+025$ the pipe is 100 m from the PVC $(11+025 - 10+925)$

curve elevation

$y = ax^2 + bx + c$

$y = -0.000076(100)^2 + 0.012(100) + 1094.3$

$y = 1094.74$ m

since the top of the pipe is at 1092.1 m, the pipe is

2.64 m below the surface

for the high pont,

$$\frac{dy}{dx} = 2ax + b = 0.000152x + 0.012 = 0$$

$x = 78.947$ m

sta $10+925 + 0+078.947 = 11+003.947$

PROBLEM 3.4

for crest curve, K at 100 km/h is 105 desirable, 62 minimum

$A = |G_2 - G_1| = 2.28$

$L_m = 105(2.28) = 239.4$ m > 150 so distance is not adequate

$L_m = 62(2.28) = 141.36$ m < 150 so distance is satisfactory

PROBLEM 3.5

$$L = 2(11+210 - 11+130) = 160 \text{ m}$$

$$G_1 = \frac{\text{elev PVC} - \text{elev PVI}}{L/2} = -\frac{322 - 320.8}{80} = -0.015$$

$$x = K|G_1|$$

$$K = \frac{x}{|G_1|} = \frac{11+190 - 11+130}{|1.5|} = 40$$

from Table 3.3, the design speed is 90 km / h desirable

PROBLEM 3.6

$$A = |G_2 - G_1| = 3.0$$

at 120 km / h, $K = 202$ for crest desirable (Table 3.2)

$$L = KA = 202(3) = 606 \text{ m}$$

f goes from 0.28 to 0.392 (1.4×0.28), and reaction time goes from 2.5 to 3 s (2.5×1.2).
Using Eq. 3.12

$$SSD = \frac{V_1^2}{2g(f \pm G)} + V_1 t_r = \frac{(120 \times 0.2778)^2}{2(9.807)(0.392)} + (120 \times 0.2778)3.0 = 244.544 \text{ m}$$

assuming $L > S$ Eq 3.14 gives

$$L_m = \frac{AS^2}{200(\sqrt{H_1} + \sqrt{H_2})^2} = \frac{3(244.54)^2}{200(\sqrt{0.9} + \sqrt{0.1})^2} = 560.64 \text{ m}$$

the 2030 curve is 45.35 m shorter

PROBLEM 3.7

$$Y = \frac{A}{200L} x^2$$

$$A = \frac{(1)(200)(245)}{(107.29)^2} = 4.26$$

$$K \text{ existing} = \frac{L}{A} = \frac{245}{4.26} = 57.51$$

which is below both minimum (62) and desirable (105) as shown in Table 3.2

acceptable length for minimum is $L = (62)(4.26) = 264.12$ m

acceptable length for desirable is $L = (105)(4.26) = 447.3$ m

PROBLEM 3.8

$K_c = 105$ (Table 3.2) 100 km/h desirable

$K_s = 51$ (Table 3.3) 100 km/h desirable

$A = |0 - 2| = 2$

$L_c = 105(2) = 210$ m

$L_s = 51(2) = 102$ m

for crest curve,

sta $PVT_c = 0+000 + 0+210 = 0+210$

elev $PVT_c = 44 - \dfrac{AL_c}{200} = 44 - \dfrac{(2)(210)}{200} = 41.9$ m

for sag curve,

sta $PVT_s = \dfrac{L_c}{2} + 1+300 + \dfrac{L_s}{2} = 1+456$

elev $PVT_s = 44 - 0.02(1300) = 18$ m

sta $PVC_s = 1+456 - 0+102 = 1+354$

elev $PVC_s = 18 + \dfrac{AL_s}{200} = 18 + \dfrac{(2)(102)}{200} = 19.02$ m

PROBLEM 3.9

$K_c = 105$ (Table 3.2) 100 km/h desirable

$K_s = 51$ (Table 3.3) 100 km/h desirable

drop in elevation is $0.02(1300) = 26$ m

$$26 = \frac{AL_c}{200} + \frac{AL_s}{200} + \frac{A(1300 - L_c - L_s)}{100}$$

substituting $K's$

$$26 = \frac{105A^2}{200} + \frac{51A^2}{200} + 13A - 1.05A^2 - 0.51A^2$$

$A = 2.32$ (the smaller of the roots)

$PVC_c = $ sta $0+000$; elev 44 m

PVT_c sta $= 0+000 + 105(2.32) = 0+243.6$

PVT_c elev $= 44 - \dfrac{AL_c}{200} = 44 - 2.826 = 41.174$ m

PVC_s elev $= 41.174 - \dfrac{A(1300 - L_c - L_s)}{100} = 41.174 - 21.763 = 19.411$ m

PVC_s sta $= 0+243.6 - (1300 - L_c - L_s) = 1+181.68$

PVT_s elev $= 19.411 - \dfrac{AL_s}{200} = 18.065$ m

PVT_s sta $= 1+181.68 + 51(2.32) = 13+00$

PROBLEM 3.10

From Table 3.2, $K = 105$

the high point is

$x = K|G_1| = 105(4) = 420$ m

at the high point

$\dfrac{dy}{dx} = 2ax + b = 2a(420) + 0.04 = 0$

$a = -0.0000476$

$y = ax^2 + bx + c$, with $c = 0$

$y = -0.0000476(420)^2 + 0.04(420)$

$y = 8.403$ m

PROBLEM 3.11

$A = 1.0 + 0.5 = 1.5$

$L = 2(15+150.2 - 14+890) = 520.4$ m

$K = \dfrac{L}{A} = \dfrac{520.4}{1.5} = 346.93$

At 100 km / h we need $K = 480$ (Table 3.4), since $346.93 < 480$ the curve is not long enough.
The curve would have to be

$L = KA = 480(1.5) = 720$ m

PROBLEM 3.12

$7.5 = \dfrac{AL_c}{200} + \dfrac{AL_s}{200}$

$K_c = 49; K_s = 32$

$\dfrac{49 A^2}{200} + \dfrac{32 A^2}{200} = 7.5$

$A = 4.303\%$

$L_c = 49(4.303) = 210.847$ m

$L_s = 32(4.303) = 137.696$ m

so total distance

$2(210.847 + 137.696) + 70 = 767.086$ m

PROBLEM 3.13

at 60 km / h, $K_c = 18$, $K_s = 18$; and $L = 1+322 - 0+122 = 1200$ m

$\Delta \text{elev} = 13$ m

$Y_{fc} + Y_{fs} + \Delta x = \Delta \text{Elev} + \Delta G_c$

$\dfrac{(4+x)^2 18}{200} + \dfrac{x^2 18}{200} + \dfrac{x(1200 - (4+x)18 - 18x)}{100} = 13 + 0.04((4+x)18)$

$x = 1.308\%$

$L_c = 18(5.308) = 95.544$ m

$L_s = 18(1.308) = 23.544$ m

$L_{constant} = 1080.912$ m

PROBLEM 3.14

at 80 km / h, $K_c = 49$

$$A = \frac{L}{K_c} = \frac{400}{49} = 8.16$$

S at 100 km / h is 205 m

with $L > S$ (Eq 3.14)

$$L_m = \frac{AS^2}{200\left(\sqrt{H_1} + \sqrt{H_2}\right)^2} = 400 = \frac{8.16(205)^2}{200\left(\sqrt{H_1} + \sqrt{0.15}\right)^2}$$

$$\sqrt{H_1} + \sqrt{0.15} = 2.07$$

$$H_1 = 2.83 \text{ m}$$

PROBLEM 3.15

at 90 km / h, $K_c = 71$, $K_s = 40$;

$$L_c = 8(71) = 568 \text{ m}$$

$$L_s = 7(40) = 280 \text{ m}$$

$$Y_{fc} = \frac{A_c L}{200} = \frac{8(568)}{200} = 22.72 \text{ m}$$

$$Y_{fs} = \frac{A_s L}{200} = \frac{7(280)}{200} = 9.8 \text{ m}$$

length for $\Delta x = (3+090 - 2+000) - 848 = 242$ m

$$\Delta x = 242(0.05) = 12.1 \text{ m}$$

$$\Delta G_c = 0.03(568) = 17.04 \text{ m}$$

$$\Delta G_s = 0.02(280) = 5.6 \text{ m}$$

$$\Delta \text{Elev} = Y_{fc} + Y_{fs} + \Delta x - \Delta G_c - \Delta G_s$$

$$\Delta \text{Elev} = 22.72 + 9.8 + 12.1 - 17.04 - 5.6$$

$$\Delta \text{Elev} = 21.98 \text{ m}$$

PROBLEM 3.16

at 60 km / h SSD = 84.6 m

at 70 km / h SSD = 110.8 m

M_s = 9 - 1.8 = 7.2 m

try 70 km / h; f_s = 0.14

$$R_v = \frac{V^2}{g\left(f_s + \dfrac{e}{100}\right)} = \frac{(70 \times 0.2778)^2}{9.807(0.22)} = 175.268 \text{ m}$$

$$M_s = R_v\left(1 - \cos\frac{90 \text{ SSD}}{\pi R_v}\right) = 175.268\left(1 - \cos\frac{90\,(110.8)}{\pi(175.26)}\right) = 8.68 \text{ m} \ \therefore \ \text{no good}$$

try 60 km / h; f_s = 0.15

$$R_v = \frac{V^2}{g\left(f_s + \dfrac{e}{100}\right)} = \frac{(60 \times 0.2778)^2}{9.807(0.23)} = 123.169 \text{ m}$$

$$M_s = R_v\left(1 - \cos\frac{90 \text{ SSD}}{\pi R_v}\right) = 123.169\left(1 - \cos\frac{90\,(84.6)}{\pi(123.169)}\right) = 7.195 \text{ m} \ \therefore \ \text{good}$$

use a 60 km / h design speed

PROBLEM 3.17

$$R_v = \frac{V^2}{g\left(f_s + \dfrac{e}{100}\right)} = \frac{(120 \times 0.2778)^2}{9.807(0.15)} = 755.439 \text{ m}$$

T = 1+568.7 - 1+346.2 = 222.5 m

$$T = R \tan\frac{\Delta}{2} = 222.5 = 755.438 \tan\frac{\Delta}{2}$$

Δ = 32.82

$$L = \frac{\pi(755.439)(32.82)}{180} = 432.728 \text{ m}$$

PT sta 1+346.2 + 0+432.728 = 1+778.928

PROBLEM 3.18

sta $PC = 7+270 - 0+155 = 7+115$

$$R = \frac{T}{\tan \dfrac{\Delta}{2}} = \frac{155}{\tan 20} = 425.859 \text{ m}$$

$$L = \frac{\pi R \Delta}{180} = \frac{\pi 425.859(40)}{180} = 297.306 \text{ m}$$

sta $PT = 7+115 - 0+297.306 = 7+412.306$

for design speed R_v is $425.859 - 4.5 = 421.359$ m

$$V = \sqrt{gR_v\left(f_s + \frac{e}{100}\right)} = \sqrt{9.807(421.359)(0.17)} = 26.5 \text{ m/s or } 95.4 \text{ km/h}$$

PROBLEM 3.19

$$\frac{e}{100} = \frac{V^2}{gR_v} - f_s = \frac{(120 \times 0.2778)^2}{9.807(300)} - 0.09 = 0.288 \text{ which exceeds allowable maximums}$$

PROBLEM 3.20

rearranging Eq 3.39

$$\tan \alpha = \frac{\dfrac{V^2}{gR_v} - f_s}{1 + f_s \dfrac{V^2}{gR_v}} = \frac{\dfrac{(180 \times 0.2778)^2}{9.807(320)} - 0.20}{1 + 0.20\left(\dfrac{(180 \times 0.2778)^2}{9.807(320)}\right)} = 0.515 \text{ m/m}$$

$$L = \frac{\pi R \Delta}{180} = \frac{\pi(320)(30)}{180} = 167.552 \text{ m}$$

$$T = R \tan \frac{\Delta}{2} = 320 \tan \frac{30}{2} = 85.744 \text{ m}$$

sta $PC = 11+511.2 - 0+085.744 = 11+425.456$

sta $PT = 11+425.456 - 0+167.552 = 11+593.008$

PROBLEM 3.21

at 110 km / h; $f_s = 0.11$

$$R_v = \frac{V^2}{g\left(f_s + \dfrac{e}{100}\right)} = \frac{(110 \times 0.2778)^2}{9.807(0.19)} = 501.141 \text{ m}$$

radius to centerline is $501.141 + 1.8 = 502.941$ m

$$T = R \tan \frac{\Delta}{2} = 502.941 \tan 17.5 = 158.577 \text{ m}$$

$$L = \frac{\pi(502.941)(35)}{180} = 307.229 \text{ m}$$

PC sta $1 + 134 - 0 + 158.577 = 0 + 975.423$

PT sta $0 + 975.423 + 0 + 307.229 = 1 + 282.652$

PROBLEM 3.22

at 120 km / h; $f_s = 0.09$

$$R_v = \frac{V^2}{g\left(f_s + \dfrac{e}{100}\right)} = \frac{(120 \times 0.2778)^2}{9.807(0.15)} = 755.439 \text{ m}$$

radius to centerline is $755.439 + 1.5 = 756.939$ m

$$L = \frac{\pi(756.939)(40)}{180} = 528.44 \text{ m}$$

PROBLEM 3.23

from Eq 3.35 $R_v = R = \dfrac{180L}{\pi\Delta} = \dfrac{180(200)}{\pi 90} = 127.324$ m

from Eq 3.40 with $M_s = 6.95$ m

$$\text{SSD} = \frac{\pi R_v}{90}\left[\cos^{-1}\left(\frac{R_v - M_s}{R_v}\right)\right] = \frac{\pi(127.324)}{90}\left[\cos^{-1}\left(\frac{127.324 - 6.95}{127.324}\right)\right] = 84.526 \text{ m}$$

looking at Table 3.1, the design speed is about 60 km / h desirable

PROBLEM 3.24

from Problem 3.21 $R_v = 501.141$ m

from Table 3.1 at 110 km/h

SSD = 179.5 m (minimum)

SSD = 246.4 m (desirable)

for mimimum using Eq 3.39

$$M_s = R_v\left(1 - \cos\frac{90 \text{ SSD}}{\pi R_v}\right) = 501.141\left(1 - \cos\frac{90(179.5)}{\pi(501.141)}\right) = 8.015 \text{ m}$$

or 8.015 - 1.8 = 6.215 m from the inside lane

for desirable using Eq 3.39

$$M_s = R_v\left(1 - \cos\frac{90 \text{ SSD}}{\pi R_v}\right) = 501.141\left(1 - \cos\frac{90(246.4)}{\pi(501.141)}\right) = 15.068 \text{ m}$$

or 15.068 - 1.8 = 13.268 m from the inside lane

SOLUTIONS: CHAPTER 4

PROBLEM 4.1

$$a = \sqrt{\frac{P}{0.1p\pi}} = \sqrt{\frac{22200}{0.1(706.8)\pi}} = 10 \text{ cm}$$

at $z = 0$ and $r = 2$ cm

$$\frac{z}{a} = 0 \text{ and } \frac{r}{a} = \frac{2}{10} = 0.2$$

from Table 4.1 $A = 1.0$ and $H = 1.97987$. Solving for Poisson's ratio using Eq 4.6

$$\Delta_z = \frac{p(1 + \mu)a}{E}\left[\frac{z}{a}A + (1 - \mu)H\right]$$

$$0.041 = \frac{706.8(1 + \mu)10}{300000}\left[0 + (1 - \mu)1.97987\right]$$

$$\mu = 0.3479$$

function $C = 0$ and function $F = 0.5$

$$\sigma_r = p\left[2\mu A + C + (1 - 2\mu)F\right]$$

$$\sigma_r = 706.8\left[2(0.3479)1.0 + 0 + (1 - 2(0.3479))0.5\right] = 599.30 \text{ kPa}$$

PROBLEM 4.2

$$530.7 = \pi a^2$$

$$a = 13.00 \text{ cm}$$

$$a^2 = \frac{P}{0.1p\pi} \quad 169 = \frac{30000}{0.1p\pi}$$

$$p = 565.33 \text{ kPa}$$

$$\frac{z}{a} = \frac{5}{13} = 0.3846 \text{ and } \frac{r}{a} = \frac{0}{13} = 0$$

from Table 4.1 $A = 0.6416$ and $H = 1.3747$ using linear interpolation.

using Eq 4.6

$$\Delta_z = \frac{p(1 + \mu)a}{E}\left[\frac{z}{a}A + (1 - \mu)H\right]$$

$$0.0855 = \frac{565.33(1 + 0.5)13}{E}\left[0.3846(0.6416) + (1 - 0.5)1.3747\right]$$

$$E = 120,440.58 \text{ kPa}$$

PROBLEM 4.3

$\dfrac{z}{a} = 0$ and $\dfrac{r}{a} = 0$

from Table 4.1 $A = 1.0,\ C = 0,\ F = 0.5,$ and $H = 2.0$

$\sigma_r = p\left[2\mu A + C + (1 - 2\mu)F\right] = 600 = p\left[2\mu + 0.5 - \mu\right]$

$p = \dfrac{600}{2\mu + 0.5 - \mu} = \dfrac{600}{\mu + 0.5}$

using Eq 4.6

$\Delta_z = \dfrac{p(1 + \mu)a}{E}\left[\dfrac{z}{a}A + (1 - \mu)H\right]$

$0.042 = \dfrac{\dfrac{600}{\mu + 0.5}(1 + \mu)9}{300000}\left[0 + (1 - \mu)2.0\right]$

$\mu = 0.287$

$p = \dfrac{600}{0.287 + 0.5} = 762.39 \text{ kPa}$

$a = \sqrt{\dfrac{P}{0.1p\pi}}$

$P = a^2(0.1p\pi) = a^2(0.1(762.39)\pi) = 19.39 \text{ kN}$

PROBLEM 4.4

$\dfrac{z}{a} = \dfrac{10}{5} = 2$ and $\dfrac{r}{a} = \dfrac{20}{5} = 4$

from Table 4.1 $A = 0.0116,\ B = -0.0041,\ C = 0.01527,\ F = -0.005465,$

and $H = 0.22418$

$\sigma_z = p\left[A + B\right] = 700[0.0116 - 0.0041] = 5.25 \text{ kPa}$

$\sigma_r = p\left[2\mu A + C + (1 - 2\mu)F\right] = 700\left[2(0.4)0.0116 + 0.01527 + (1 - 0.8)(-0.005465)\right]$

$\quad = 16.42 \text{ kPa}$

$\Delta_z = \dfrac{p(1 + \mu)a}{E}\left[\dfrac{z}{a}A + (1 - \mu)H\right] = \dfrac{700(1 + 0.4)5}{250000}\left[2(0.0116) + (1 - 0.4)0.22418\right]$

$\quad = 0.00309 \text{ cm}$

PROBLEM 4.5

From Eq 4.9

SN = 0.44(3) + 0.2(6)(1) + 0.11(8)(1) = 3.4

using linear interpolation for Truck A

single 12-kip = 0.229 -0.4(0.016) = 0.2226, single 23-kip = 2.63 -0.4(0.14) = 2.574. So total 18-kip ESAL = 2.7966

for Truck B

single 8-kip = 0.005 -0.4(0.001) = 0.0046, tandem 43-kip = 2.74 -0.4(0.085) = 2.706. So total 18-kip ESAL = 2.7106

Therefore, Truck A causes more damage.

PROBLEM 4.6

From Eq 4.9

SN = 0.44(4) + 0.18(7)(0.9) + 0.11(10)(0.8) = 3.774

Z_R = -1.282 at 90% (Table 4.5)

applying Eq. 4.7 gives W_{18} = 942,046.45 18-kip ESAL

for 18-kip ESAL, interpolating: 3.7 - 0.774(0.3) = 3.4678

so 942046.45/3.4678 = 271,655.36 18-kip loads

PROBLEM 4.7

From Eq 4.9

SN = 0.35(4) + 0.20(6)(1) + 0.11(7)(1) = 3.37

axle loads (interpolating)

single 10-kip	0.1121 × 300	= 33.624
single 18-kip	1 × 120	= 120.0
single 23-kip	2.578 × 100	= 257.80
tandem 32-kip	0.8886 × 100	= 88.86
single 32-kip	9.871 × 30	= 296.13
triple 40-kip	0.546 × 100	= 54.623

which is a total of 851.037

W_{18} = 851.037 × 10 × 365 = 3,106285.05

Z_R = -1.036 (Table 4.5)

ΔPSI = 4.7 - 2.5 = 2.2

applying Eq. 4.7 gives M_R = 9009.87 psi

PROBLEM 4.8

$SN = 0.35(4) + 0.20(6)(1) + 0.11(7)(1) = 3.37$

axle loads (interpolating)

single 10-kip	0.1121×300	$= 33.624$
single 18-kip	1×20	$= 20.0$
single 23-kip	2.682×100	$= 268.18$
tandem 32-kip	0.8886×100	$= 88.86$
single 32-kip	9.871×90	$= 888.39$
triple 40-kip	0.546×100	$= 54.623$

which is a total of 1353.677

$W_{18} = 1353.677 \times 10 \times 365 = 4,940,921.05$

applying Eq. 4.7 gives $M_R = 11,005$ psi

PROBLEM 4.9

From Eq 4.9

$SN = 0.44(4) + 0.40(4)(1) + 0.11(8)(1) = 4.24$

axle loads (interpolating)

single 8-kip	0.40832×1300	$= 530.82$
tandem 15-kip	0.0431×900	$= 38.77$
single 40-kip	22.164×20	$= 443.28$
tandem 40-kip	2.042×200	$= 408.4$

which is a total of 1421.27

$W_{18} = 1421.27 \times 12 \times 365 = 6,225,126.6$

$Z_R = -0.524$ (Table 4.5)

applying Eq. 4.7 gives $M_R = 6269.96$ psi

PROBLEM 4.10

From Eq 4.9

$SN = 0.35(6) + 0.20(9)(1) + 0.11(10)(1) = 5.0$

axle loads

single 2-kip	0.0004×20000	$= 8$
single 10-kip	0.088×200	$= 17.6$
tandem 22-kip	0.18×200	$= 36.0$
single 12-kip	0.189×410	$= 77.49$

tandem 18-kip 0.077×410 $= 31.57$

triple 50-kip 1.22×410 $= 500.2$

which is a total of 670.86

$Z_R = -1.645$ (Table 4.5)

applying Eq. 4.7 gives $W_{18} = 3,545,632.84$

years $= 3,545,632.84/(670.86 \times 365) = 14.48$

PROBLEM 4.11

assume SN = 4.0

axle loads

single 10-kip 0.102×5000 $= 510$

single 24-kip 2.89×400 $= 1156$

tandem 30-kip 0.695×1000 $= 695$

tandem 50-kip 4.64×100 $= 464$

which is a total of 2825

directional $W_{18} = 2825 \times 365 \times 15 = 15,466,875$

$PDL = 0.8$ (Table 4.11)

design lane $W_{18} = 0.8 \times 15,466,875 = 12,373,500$

$Z_R = -1.282$ (Table 4.5)

applying Eq. 4.7 gives SN = 3.996 (assumed value of 4 is good)

PROBLEM 4.12

From Eq 4.9

$SN = 0.44(5) + 0.40(6)(1) + 0.11(10)(1) = 5.7$

axle loads (interpolating)

single 20-kip 1.538×200 $= 307.6$

tandem 40-kip 2.122×200 $= 424.4$

single 22-kip 2.264×80 $= 181.12$

which is a total of 913.12

$W_{18} = 913.12 \times 20 \times 365 = 6,665,776$

applying Eq. 4.7 gives $Z_R = -0.6993$

from Table 4.5, $R = 75.78\%$

PROBLEM 4.13

From Eq 4.9

$SN = 0.35(4) + 0.18(6)(1) + 0.11(8)(1) = 3.36$

$W_{18} = 1290 \times 20 \times 365 = 9{,}417{,}000$

applying Eq. 4.7 gives $Z_R = -0.1611$

from Table 4.5, $R = 56.404\%$

PROBLEM 4.14

$$a = \sqrt{\frac{P}{p\pi}} = \sqrt{\frac{10000}{90\pi}} = 5.95 \text{ in}$$

$$\sigma_e = 0.529(1 + 0.54\mu)\left(\frac{P}{h^2}\right)\left[\log_{10}\left(\frac{Eh^3}{ka^4}\right) - 0.71\right]$$

$$218.5 = 0.529(1 + 0.54(0.25))\left(\frac{10000}{h^2}\right)\left[\log_{10}\left(\frac{4200000h^3}{150(5.95)^4}\right) - 0.71\right]$$

$$h = 10.00 \text{ in}$$

PROBLEM 4.15

$$l = \left(\frac{Eh^3}{12(1 - \mu^2)k}\right)^{0.25} = 30.106 = \left(\frac{3500000(8)^3}{12(1 - (0.3)^2)k}\right)^{0.25}$$

$$k = 199.75 \text{ pci}$$

$$\Delta_i = \frac{P}{8kl^2}\left(1 + \left(\frac{1}{2\pi}\right)\left(\ln\left(\frac{a}{2l}\right) + \gamma - \frac{5}{4}\right)\left(\frac{a}{l}\right)^2\right)$$

$$0.008195 = \frac{12000}{8(199.75)(30.106)^2}\left(1 + \left(\frac{1}{2\pi}\right)\left(\ln\left(\frac{a}{2(30.106)}\right) + 0.577215 - \frac{5}{4}\right)\left(\frac{a}{(30.106)}\right)^2\right)$$

$$a = 4.33 \text{ in.}$$

$$\sigma_i = \frac{3P(1 + \mu)}{2\pi h^2}\left(\ln\left(\frac{2l}{a}\right) + 0.5 - \gamma\right) + \frac{3P(1 + \mu)}{64h^2}\left(\frac{a}{l}\right)^2$$

$$\sigma_i = \frac{3(12000)(1 + 0.3)}{2\pi(8)^2}\left(\ln\left(\frac{2(30.106)}{4.33}\right) + 0.5 - 0.577215\right) + \frac{3(12000)(1 + 0.3)}{64(8)^2}\left(\frac{4.33}{(30.106)}\right)^2$$

$$\sigma_i = 297.66 \text{ psi}$$

PROBLEM 4.16

$$\Delta_c = \frac{P}{kl^2}\left[1.205 - 0.69\left(\frac{a_l}{l}\right)\right] = 0.05 = \frac{17000}{250(10)^2}\left[1.205 - 0.69\left(\frac{7}{l}\right)\right]$$

$l = 38.306$

$$l = \left(\frac{Eh^3}{12(1 - \mu^2)k}\right)^{0.25} = 35.496 = \left(\frac{E(10)^3}{12(1 - (0.36)^2)(250)}\right)^{0.25}$$

$E = 5,621,954.06$ psi

PROBLEM 4.17

$$l = \left(\frac{Eh^3}{12(1 - \mu^2)k}\right)^{0.25} = \left(\frac{4000000(12)^3}{12(1 - (0.4)^2)300}\right)^{0.25} = 38.88$$

$$\sigma_i = \frac{3P(1 + \mu)}{2\pi h^2}\left(\ln\left(\frac{2l}{a}\right) + 0.5 - \gamma\right) + \frac{3P(1 + \mu)}{64h^2}\left(\frac{a}{l}\right)^2$$

$$\sigma_i = \frac{3(9000)(1 + 0.4)}{2\pi(12)^2}\left(\ln\left(\frac{2(38.88)}{5}\right) + 0.5 - 0.577215\right) + \frac{3(9000)(1 + 0.4)}{64(12)^2}\left(\frac{5}{(38.883)}\right)^2$$

$\sigma_i = 111.51$ psi

$$\Delta_i = \frac{P}{8kl^2}\left(1 + \left(\frac{1}{2\pi}\right)\left(\ln\left(\frac{a}{2l}\right) + \gamma - \frac{5}{4}\right)\left(\frac{a}{l}\right)^2\right)$$

$$\Delta_i = \frac{9000}{8(300)(38.88)^2}\left(1 + \left(\frac{1}{2\pi}\right)\left(\ln\left(\frac{5}{2(38.88)}\right) + 0.577215 - \frac{5}{4}\right)\left(\frac{5}{(38.88)}\right)^2\right) = 0.00328 \text{ in.}$$

$$\sigma_e = 0.529(1 + 0.54\mu)\left(\frac{P}{h^2}\right)\left[\log_{10}\left(\frac{Eh^3}{ka^4}\right) - 0.71\right]$$

$$\sigma_e = 0.529(1 + 0.54(0.4))\left(\frac{9000}{(12)^2}\right)\left[\log_{10}\left(\frac{4000000(12)^3}{300(5)^4}\right) - 0.71\right] = 155.05 \text{ psi}$$

$$\Delta_e = 0.408(1 + 0.4\mu)\left(\frac{P}{kl^2}\right) = 0.408(1 + 0.4(0.4))\left(\frac{9000}{300(38.88)^2}\right) = 0.00939 \text{ in.}$$

PROBLEM 4.18

Truck A

single 12-kip = 0.175
single 23-kip = 2.915

total 18-kip ESAL = 3.09

for Truck B

single 8-kip = 0.032
tandem 23-kip = 5.245

total 18-kip ESAL = 5.277

Therefore, Truck B causes more damage.

PROBLEM 4.19

Assume $D = 10$

axle loads

single 22.5-kip	2.648×80	$= 211.8$
tandem 25-kip	0.5305×570	$= 302.39$
tandem 39-kip	3.495×50	$= 174.75$
triple 48-kip	2.55×80	$= 204.0$

which is a total of 892.94

$Z_R = -1.645$ (Table 4.5)

$W_{18} = 892.94 \times 20 \times 365 = 6,518,462$

applying Eq. 4.18 gives $D = 10.79$ in. which is close to the assumed 10 in.

PROBLEM 4.20

axle loads (interpolating)

single 8-kip	0.0325×1300	$= 42.25$
tandem 15-kip	0.06575×900	$= 59.175$
single 40-kip	26×80	$= 520.0$
tandem 40-kip	3.645×200	$= 729.0$

which is a total of 1350.425

$Z_R = -0.524$ (Table 4.5)

$W_{18} = 1350.425 \times 12 \times 365 = 5,914,861.5$

applying Eq. 4.18 gives $S'_c = 575.34$ psi

PROBLEM 4.21

for rigid pavement $D = 10$ in, axle loads

single 20-kip	1.58×100	$= 158$
tandem 42-kip	4.74×100	$= 474$

which is a total of 632

for flexible pavement SN = 4, axle loads

single 20-kip	1.47×100	$= 147$
tandem 42-kip	2.43×100	$= 243$

which is a total of 390

$Z_R = -1.282$ (Table 4.5)

for rigid, applying Eq. 4.18 gives $W_{18} = 1,472,351.22$

years $= 1,472,351.22/(632 \times 365) = 6.38$

for flexible pavement, applying Eq. 4.7 with $W_{18} = 908,193$ ($15.07 \times 390 \times 365$) gives

$M_R = 3668.88$ psi

PROBLEM 4.22

with $D = 8$ in, axle loads

single 10-kip	0.084×300	$= 25.2$
single 18-kip	1.00×200	$= 200$
single 23-kip	2.75×100	$= 275$
tandem 32-kip	1.47×100	$= 147$
single 32-kip	10.1×30	$= 303$
tandem 40-kip	1.16×100	$= 116$

which is a total of 1066.2

$Z_R = -1.036$ (Table 4.5)

applying Eq. 4.18 gives $W_{18} = 2,883,385$

years $= 2,883,385/(1066.2 \times 365) = 7.41$

PROBLEM 4.23

with $D = 8$ in, axle loads

single 10-kip	0.084×300	$= 25.2$
single 18-kip	1.00×200	$= 200$
single 23-kip	2.75×100	$= 275$
tandem 32-kip	1.47×100	$= 147$
single 32-kip	10.1×30	$= 303$
tandem 40-kip	1.16×100	$= 116$

which is a total of 1066.2

$Z_R = -1.645$ (Table 4.5)

applying Eq. 4.18 gives $W_{18} = 1,893,227$

years $= 1,893,227/(1066.2 \times 365) = 4.86$

PROBLEM 4.24

from Example 4.5 daily $W_{18} = 1449.27$

applying Eq. 4.18 with $C_d = 0.8$ gives $W_{18} = 8,148,507.94$

years $= 8,148,507.94/(1449.27 \times 365) = 15.40$

applying Eq. 4.18 with $C_d = 0.6$ gives $W_{18} = 3,046,404.82$

years $= 3,046,404.82/(1449.27 \times 365) = 5.76$

PROBLEM 4.25

using conditions from Example 4.6 and applying Eq. 4.18 gives $W_{18} = 37,684,418$

years $= 37,684,418/(4.434 \times 1227.76 \times 365) = 18.965$

a 1.035 year reduction

PROBLEM 4.26

using conditions from Example 4.6 and applying Eq. 4.18 gives $W_{18} = 14,857,329.8$

years $= 14,857,329.8/(4.434 \times 1227.76 \times 365) = 7.477$

a 12.523 year reduction

PROBLEM 4.27

$$Q = VA_c$$

$$= \frac{Q}{A_c}$$

$$= \frac{R_h^{2/3} S^{1/2}}{n} = \frac{Q}{A_c} = \frac{R_h^{2/3} S^{1/2}}{n}$$

using area and hydraulic radius values from Table 4.16

$$Q = \left(\frac{2}{3}(1.6)d\right)\left(\frac{2d(1.6)}{3(1.6)^2 + 8d^2}\right)^{2/3} (0.035)^{1/2} \Big/ n$$

$$d = 0.80 \text{ m}$$

PROBLEM 4.28

as in problem 4.27,

$$Q = \left((1.6)d\right)\left(\frac{d(1.6)}{(1.6) + 2d}\right)^{2/3} (0.035)^{1/2} \Big/ n$$

$$d = 0.369 \text{ m}$$

PROBLEM 4.29

using 25-yr return period and rainfall duration = 30 min, $I = 69$ mm/h

$C_r = 0.8$ (Table 4.12, conservative)

$$Q = 0.00278(C_r I A_d) = 0.00278(0.8)(69)(24) = 3.68 \text{ m}^3/\text{s}$$

for culvert,

$$Q = VA_c$$

$A_c = 1.84$ m² (i.e., 3.68/2)

so dimensions 1.36 m sides

31

PROBLEM 4.30

from Table 4.16, with b $=$ 2 and z $=$ 1,

$$R_h = \frac{2d + 8d^2}{b + 2d\sqrt{z^2 + 1}} = \frac{2d + 8d^2}{2 + 2.828d}$$

from Table 4.15 $n = 0.015$

$$= \frac{R_h^{2/3} S^{1/2}}{n} = \frac{\left(\dfrac{2d + 8d^2}{2 + 2.828d}\right)^{2/3} (0.025)^{1/2}}{0.015} = 10$$

solving gives d $=$ 1.7 m

PROBLEM 4.31

from Table 4.12 $C_r = 0.70$

from Table 4.13 years $= 50$

$A_c = \pi r^2 = 3.14(0.5)^2 = 0.785$ m²

$Q = VA_c = 0.00278(C_r I A_d)$

$2(0.785) = 0.00278(0.7)(17) I$

$I = 47.46$, from Fig. 4.12, the time of concentration is about 60 min.

PROBLEM 4.32

$I = 44$ mm/h for a 50-yr return period and 70 min. duration (Fig. 4.12)

$Q = VA_c = 0.00278(C_r I A_d)$

$2(A_c) = 0.00278(0.7)(44)(29)$

$A_c = 1.2415$

$$d = 2\sqrt{\frac{1.2415}{3.14}} = 1.258 \text{ m}$$

use a 1.5 m diameter

PROBLEM 5.1

$$u = u_f \left[1 - \left(\frac{k}{k_j} \right)^{3.5} \right]$$

and $q = ku = u_f \left[k - k \left(\frac{k}{k_j} \right)^{3.5} \right]$

at capacity $\frac{dq}{dk} = 0 = \left[1 - 4.5 \left(\frac{k_m^{3.5}}{k_j^{3.5}} \right) \right] = \left[1 - 4.5 \left(\frac{k_m^{3.5}}{(140)^{3.5}} \right) \right]$

$k_m = 91.09 \text{ veh / km}$

$q_m = k_m u_m$

$u_m = \frac{3800}{91.09} = 41.72 \text{ km / h}$

$u_f = \frac{3800}{\left[k_m - k_m \left(\frac{k_m}{k_j} \right)^{3.5} \right]} = \frac{3800}{\left[91.09 - (91.09) \left(\frac{91.09}{140} \right)^{3.5} \right]} = 53.63 \text{ km / h}$

PROBLEM 5.2

at maximum flow $q = 2900$ veh/h and $u = 50$ km/h and,

$\frac{dq}{du} = 2au + b = 0$

$b = -100a$

at $q = 2900$ and $u = 50$, $q = a(50)^2 + b(50) = 2900$

$2900 = 2500a + 50b$

substituting: $2900 = 2500a - 5000a$

so $a = -1.16$ and $b = 116$

at $q = 1400$ with $q = -1.16(u)^2 + 116(u)$; $u = 14.04$ km/h or 85.96 km/h

at $q = 0$ with $q = -1.16(u)^2 + 116(u)$; $u = 100$ km/h, the free-flow speed

PROBLEM 5.3

$$q = 80k - 0.4k^2$$

$$\frac{dq}{dk} = -0.8k_m + 80 = 0$$

$$k_m = 100 \text{ veh / km}$$

$$q_m = 80(100) - 0.4(100)^2 = 4000 \text{ veh / h}$$

$$u_m = \frac{4000}{100} = 40 \text{ km / h}$$

at 0.25 of capacity $q = 1000 \text{ veh / h}$ so,

$$80k - 0.4k^2 - 1000 = 0$$

$$k = 186.6 \text{ veh / km or } 13.40 \text{ veh / km}$$

PROBLEM 5.4

$$P(h \geq 1.3) = e^{-q(13)/3600}$$

$$0.6 = e^{-0.0036q}$$

$$q = 141.9 \text{ veh / h or } 0.0394 \text{ veh / s}$$

during 30 s intervals, $30 \times 0.0394 = 1.18 \text{ veh}$

$$P(4) = \frac{(1.18)^4 e^{-1.18}}{4!} = 0.025$$

PROBLEM 5.5

$$P(0) = \frac{(x)^0 e^{-x}}{0!} = 0.15$$

$$x = 1.90 \text{ veh / interval}$$

$$P(3) = \frac{(1.90)^3 e^{-1.90}}{3!} = 0.1715 \text{ and intervals} = 20.58 \ (0.1715 \times 120)$$

PROBLEM 5.6

$$\text{veh}/\text{s} = \frac{1.9 \text{ veh}/\text{int}}{20 \text{ s}/\text{int}} = 0.095 \text{ veh}/\text{s}$$

$$P(h \geq 10) = e^{-0.095(10)} = 0.387 \text{ or } 38.7\%$$

$$P(h \geq 6) = e^{-0.095(6)} = 0.565$$

$$P(h < 6) = 1 - P(h \geq 6) = 0.435 \text{ or } 43.5\%$$

PROBLEM 5.7

want the probability that the headway is less than 4 s (1.5 + 2.5)

$$P(h \geq 4) = e^{-280(4)/3600} = 0.7326$$

$$P(h < 4) = 1 - P(h \geq 4) = 1 - 0.7326 = 0.267$$

PROBLEM 5.8

want

$$P(h < 4) = 0.015 \text{ so } P(h \geq t) = 0.85$$

$$0.85 = e^{-280t/3600}$$

$$t = 2.089 \text{ s}$$

so reaction time = 2.089 - 1.5 = 0.59 s

PROBLEM 5.9

$$\lambda(t) \quad = 6t \qquad\qquad 0 \leq t \leq 30$$
$$\quad = 180 + 2(t\text{-}30) \qquad t > 30$$

$$\mu(t) \quad = 0 \qquad\qquad 0 \leq t \leq 15$$
$$\quad = 6(t\text{-}15) \qquad\quad t > 15$$

for dissipation:

$$180 + 2(t\text{-}30) = 6(t\text{-}15)$$

$$t = 52.5 \text{ min}$$

using areas, delay is

$$\frac{30(180)}{2} - \frac{15(90)}{2} + \frac{90(52.5 - 30)}{2} = 3037.5 \text{ veh} - \text{min}$$

longest queue is 90 from 15 to 30 minutes, under FIFO the longest delay is 15 min for the first 90 cars, under LIFO the delay is 52.5 min (until queue intersection).

PROBLEM 5.10

for dissipation, after 30 min note that:

$40 + 2t = 4t$

$t = 20$ min, so dissipation is at 50 min (20 + 30).

using areas, delay is
$$\frac{20(80)}{2} + \frac{(80 + 40)(10)}{2} + \frac{40(20)}{2} = 1800 \text{ veh} - \text{min}$$

PROBLEM 5.11

$\lambda = \int 4.1 + 0.01t = 4.1t + 0.005t^2$ and $\mu = 12(t - 10)$

$4.1t + 0.005t^2 = 12(t - 10)$

$t = 15.34 \text{ min}$

number that arrive $= 4.1(15.34) + 0.005(15.34)^2 = 64.07$

number that depart $= 12(5.34) = 64.08$

for delay,

$$D_t = \int_0^{15.34} 4.1t + 0.005t^2 - \frac{1}{2}(5.34)(64.07) = 2.05t^2 + 0.00167t^3|_0^{15.34} - 171.067 = 317.358 \text{ veh} - \text{min}$$

PROBLEM 5.12

$$\rho = \frac{4}{5} = 0.8; \quad \overline{Q} = \frac{\rho^2}{2(1 - \rho)} = \frac{(0.8)^2}{2(1 - 0.8)} = 1.6 \text{ veh}$$

$$\overline{w} = \frac{\rho}{2\mu(1 - \rho)} = \frac{0.8}{2(5)(1 - 0.8)} = 0.4 \text{ min}$$

$$\overline{t} = \frac{2 - \rho}{2\mu(1 - \rho)} = \frac{2 - 0.8}{2(5)(1 - 0.8)} = 0.6 \text{ min}$$

PROBLEM 5.13

$$\rho = \frac{4}{5} = 0.8; \quad \overline{Q} = \frac{\rho^2}{1 - \rho} = \frac{(0.8)^2}{1 - 0.8} = 3.2 \text{ veh}$$

$$\overline{w} = \frac{\lambda}{\mu(\mu - \lambda)} = \frac{4}{5(5 - 4)} = 0.8 \text{ min}$$

$$\overline{t} = \frac{1}{\mu - \lambda} = \frac{1}{5 - 4} = 1 \text{ min}$$

PROBLEM 5.14

$$\rho = \lambda/\mu = 2/3$$

$$\overline{Q} = \frac{\rho^2}{2(1 - \rho)} = \frac{(2/3)^2}{2(1 - 2/3)} = 0.667 \text{ veh}$$

$$\overline{w} = \frac{\rho}{2\mu(1 - \rho)} = \frac{2/3}{2(3)(1 - 2/3)} = 0.35 \text{ min}$$

$$\overline{t} = \frac{2 - \rho}{2\mu(1 - \rho)} = \frac{2 - 2/3}{2(3)(1 - 2/3)} = 0.667 \text{ min}$$

PROBLEM 5.15

$$\lambda(t) \quad = 4t \qquad\qquad 0 \le t \le 30$$
$$= 120 + 8(t-30) \qquad 30 < t \le 75$$

$$\mu(t) \quad = 0 \qquad\qquad 0 \le t \le 30$$
$$= 11(t-30) \qquad 30 < t \le 75$$

for dissipation:

$120 + 8(t\text{-}30) = 11(t\text{-}30)$

$t = 70$ min

using areas delay is (with 120 arriving by $t = 30$ and 440 by $t = 70$ min)

$$\frac{30(120)}{2} + \frac{120(40)}{2} = 4200 \text{ veh} - \text{min}$$

longest queue is 120 at $t = 30$, under FIFO the longest delay is 30 min for the first car, under LIFO the delay is 70 min (until queue intersection).

PROBLEM 5.16

distribution rate required to clear the queue,

$\lambda(t) \quad = 120 + 8(t\text{-}30)$
$\mu(t) \quad = \mu(t\text{-}30)$

for dissipation at $t = 60$:

$120 + 8(60\text{-}30) = \mu(60\text{-}30)$

$\mu = 12$ veh/min

PROBLEM 5.17

at queue dissipation, $\lambda t = 4(t\text{-}t_s)$, where t_s is the time until processing begins

$t_s = 30/\lambda$ which is also the time that the lane is full, $t = 30$ when the queue dissipates so,

$\lambda(30) = 4(30\text{-}30/\lambda)$ which gives

$\lambda = 2$ veh/min

PROBLEM 5.18

$\lambda t = x$ and, since $\lambda = 2$, $t = x/2$, at queue dissipation,

$\lambda t = \mu(t\text{-}13)$, substituting $t = x/2$ gives : $\mu = \dfrac{x}{\dfrac{x}{2} - 13}$

38

PROBLEM 5.19

$(1/2)bh = 3600$

$(1/2)(30)h = 3600$

$h = 240$ veh, so it will take 60 min (240/4) for the queue to dissipate since $\mu = 4$

PROBLEM 5.20

time to queue dissipation is

$$\int_0^t 4.3 - 0.22t \ dt = 2t$$

$$4.3t - 0.11t^2 = 2t$$

$$t = 20.91 \text{ min}$$

for delay,

$$D_t = \int_0^{20.91} 4.3t - 0.11t^2 - \frac{1}{2}(20.91)(41.82) = 2.15t^2 - 0.0367t^3 \big|_0^{20.91} - 437.23 = 167.28 \text{ veh} - \text{min}$$

length of queue at t is

$$Q(t) = 4.3t - 0.11t^2 - 2t = 2.3t - 0.11t^2$$

for maximum

$$\frac{dQ(t)}{dt} = 0 = 0.22t - 2.3$$

$t = 10.45$ min, so $Q(10.45) = 2.3(10.45) - 0.11(10.45)^2 = 12.02 \text{ veh}$

PROBLEM 5.21

length of queue at t is

$$Q(t) = \int_0^t 3.3 - 0.10t \; dt - \mu t$$

$$Q(t) = 3.3t - 0.05t^2 - \mu t$$

for maximum

$$\frac{dQ(t)}{dt} = 0 = 3.3 - 0.10t - \mu$$

$$t = \frac{3.3 + \mu}{0.1}$$

substituting,

$$Q(t) = 3.3\left(\frac{3.3 + \mu}{0.1}\right) - 0.05\left(\frac{3.3 + \mu}{0.1}\right)^2 - \mu\left(\frac{3.3 + \mu}{0.1}\right) = 4$$

$$\mu = 2.406 \; \text{veh / min}$$

PROBLEM 5.22

queue $is \; M / M / 1$

$$\rho = \lambda/\mu = 0.75$$

$$P_0 = \cfrac{1}{\sum_{n=0}^{N-1}\cfrac{\rho^n}{n!} + \cfrac{\rho^N}{N!(1-\rho/N)}} = \cfrac{1}{1 + \cfrac{0.75}{0.25}} = 0.25$$

$$P_n = \frac{\rho^n P_0}{N^{n-N} N!}$$

$$P_1 = \frac{0.75(0.25)}{1} = 0.1875$$

$$P_2 = \frac{0.75^2 (0.25)}{1} = 0.1406$$

$$P_3 = \frac{0.75^3(0.25)}{1} = 0.105$$

$$P_4 = \frac{0.75^4(0.25)}{1} = 0.079$$

$$P_5 = \frac{0.75^5(0.25)}{1} = 0.059$$

$$P(n > 5) = 1 - 0.8214 = 0.1786$$

PROBLEM 5.23

trial and error, at 6 spaces

$$P_0 = \cfrac{1}{1 + \cfrac{2}{1!} + \cfrac{2^2}{2!} + \cfrac{2^3}{3!} + \cfrac{2^4}{4!} + \cfrac{2^5}{5!} + \cfrac{2^6}{6!(0.5)}}$$

$$= 0.134$$

$$P_{n>N} = \frac{P_0 \, \rho^{N+1}}{N! N (1 - \rho/N)}$$

$$P_{n>6} = \frac{(0.134)(2)^7}{6!6(0.5)} = 0.007 \text{ which is the answer because at 5 spaces}$$

$$P_0 = \cfrac{1}{1 + \cfrac{2}{1!} + \cfrac{2^2}{2!} + \cfrac{2^3}{3!} + \cfrac{2^4}{4!} + \cfrac{2^5}{5!(0.5)}}$$

$$= 0.1327$$

$$P_{n>N} = \frac{P_0 \, \rho^{N+1}}{N! N (1 - \rho/N)}$$

$$P_{n>5} = \frac{(0.1327)(2)^6}{5!5(0.5)} = 0.028$$

PROBLEM 5.24

$\rho = 7.167/6 = 1.1944$

$\rho/N = 0.597 < 1$ so equations apply

$$P_0 = \frac{1}{1 + \dfrac{1.1944}{1!} + \dfrac{1.1944^2}{2!(0.403)}} = 0.252$$

$$\overline{Q} = \frac{P_0 \, \rho^{N+1}}{N! \, N}\left[\frac{1}{(1-\rho/N)^2}\right] = \frac{0.252\,(1.1944)^{2+1}}{2!\,2}\left[\frac{1}{\left(1-(1.1944)/2\right)^2}\right] = 0.661$$

$$\overline{t} = \frac{\rho + \overline{Q}}{\lambda} = \frac{1.1944 + 0.661}{7.167} = 0.259$$

so total time spent in one hour:

$430(0.259) = 111.32 \text{ veh - min}$

PROBLEM 5.25

$\rho = 8.33/4 = 2.08$

N must be greater than 2 for the equations to apply, try 3

$$P_0 = \frac{1}{1 + \dfrac{2.08}{1!} + \dfrac{2.08^2}{2!} + \dfrac{2.08^3}{3!(0.307)}} = 0.099$$

$$\overline{Q} = \frac{P_0 \, \rho^{N+1}}{N! \, N}\left[\frac{1}{(1-\rho/N)^2}\right] = \frac{0.099\,(2.08)^4}{3!\,3}\left[\frac{1}{\left(1-(2.08)/3\right)^2}\right] = 1.095$$

$$\overline{w} = \frac{\rho + \overline{Q}}{\lambda} - \frac{1}{\mu} = \frac{2.08 + 1.095}{8.33} - \frac{1}{4} = 0.1311 \text{ min or } 7.87 \text{ s}$$

try 4

$$P_0 = \frac{1}{1 + \dfrac{2.08}{1!} + \dfrac{2.08^2}{2!} + \dfrac{2.08^3}{3!} + \dfrac{2.08^4}{4!(0.307)}} = 0.108$$

$$\overline{Q} = \frac{P_0 \, \rho^{N+1}}{N! \, N}\left[\frac{1}{(1-\rho/N)^2}\right] = \frac{0.108\,(2.08)^5}{4!\,4}\left[\frac{1}{\left(1-(2.08)/4\right)^2}\right] = 0.190$$

$$\overline{w} = \frac{\rho + \overline{Q}}{\lambda} - \frac{1}{\mu} = \frac{2.08 + 0.190}{8.33} - \frac{1}{4} = 0.0225 \text{ min or } 1.35 \text{ s} \therefore \text{ must have 4 booths open}$$

SOLUTIONS: CHAPTER 6

PROBLEM 6.1

$\lambda = 800/3600 = 0.222$; $\mu = 1500/3600 = 0.417$

$\rho = \lambda/\mu = 0.533$

at queue clearing, $r + t_0 + 10 = c$, so $t_0 = c - r - 10$

$$t_0 = \frac{\rho r}{(1-\rho)} = 60 - r - 10 = \frac{0.533r}{(0.467)}$$

$r = 23.36$ s

PROBLEM 6.2

from the area of a triangle $D_t = (r \times \lambda c)/2$

and $\lambda c = \mu g$

$\mu = 1000/3600 = 0.278$

$\lambda = \mu g/c$ and $g = c - r = c - 30$

substituting

$\lambda = (0.278)(c-30)/c = 0.278 - 8.34/c$

$D_t = (r \times (0.278c - 8.34))/2 = 83.33 = (30 \times (0.278c - 8.34))/2$

$c = 50$ s

$\lambda = 0.278 - 8.34/50 = 0.1112$ veh/s or 400 veh/h

arrivals $= \lambda c = 0.1112(50) = 5.56$ veh

PROBLEM 6.3

$\lambda = 500/3600 = 0.139$; $\mu = 1400/3600 = 0.389$

$\rho = \lambda/\mu = 0.139/0.389 = 0.357$

$r = c - g = 60 - 25 = 35$

$\lambda c = 8.34$; $\mu g = 9.72$

for $D/D/1$:

$$d = \frac{r^2}{2c(1-\rho)} = \frac{(35)^2}{2(60)(1-0.357)} = 15.88 \text{ s}$$

for Webster's formula $x = \lambda c / \mu g = 0.858$

$$d' = d + \frac{x^2}{2\lambda(1-x)} - 0.65\left(\frac{c}{\lambda^2}\right)^{1/3} x^{2+5(g/c)}$$

$$= 15.88 + \frac{(0.858)^2}{2(0.139)(1-0.858)} - 0.65\left(\frac{60}{(0.139)^2}\right)^{1/3}(0.858)^{2+5(25/60)} = 29.45 \text{ s}$$

for Allsop's formula

$$d' = \frac{9}{10}\left[d + \frac{x^2}{2\lambda(1-x)}\right] = \frac{9}{10}\left[15.88 + \frac{(0.858)^2}{2(0.139)(1-0.858)}\right] = 31.08 \text{ s}$$

PROBLEM 6.4

$Q = \lambda r$

$\lambda = Q/r$

$\lambda = 8/40 = 0.2$ veh/s; $\mu = 1440/3600 = 0.4$ veh/s

$\rho = \lambda/\mu = 0.2/0.4 = 0.5$

$$t_0 = \frac{\rho r}{(1-\rho)} = \frac{0.5(40)}{(0.5)} = 40 \text{ s}$$

PROBLEM 6.5

from Allsop's formula

$$d - \frac{d'}{0.9} = \frac{x^2}{2\lambda(1-x)} = \frac{16.6}{0.9} - 11.25 = \frac{(0.8)^2}{2\lambda(1-0.8)}$$

$\lambda = 0.222$ veh/s or 800 veh/h

also, $x = \lambda c/\mu g = 0.8$

$\mu = 1600/3600 = 0.444$, $c = 100$, $g = 50$

$0.222c/0.444(50) = 0.8$

$c = 80$

PROBLEM 6.6

$d' - d = 34$ or $d' = 34 + d$

$x = \lambda c / \mu g = \lambda(60) / (1300/3600)25 = 6.65\lambda$

$$d = \frac{r^2}{2c(1-\rho)} = \frac{(35)^2}{2(60)(1-\lambda/0.3611)} = \frac{1225}{120-332.32\lambda}$$

$$d' = \frac{9}{10}\left[d + \frac{x^2}{2\lambda(1-x)}\right] = \frac{9}{10}\left[\frac{1225}{120-332.32\lambda} + \frac{(6.65\lambda)^2}{2(\lambda)(1-6.65\lambda)}\right]$$

so,

$$\frac{1225}{120-332.32\lambda} - 34 = \frac{9}{10}\left[\frac{1225}{120-332.32\lambda} + \frac{(6.65\lambda)^2}{2(\lambda)(1-6.65\lambda)}\right]$$

$\lambda = 0.3705$ or 0.1387 (quadratic)

since 0.3705 veh / s $= 1333.8$ veh / h (which is greater than 1300 veh / h) the solution must be 0.1387 veh / s or 500 veh / h

PROBLEM 6.7

for delay, a graph of the queue gives the area between the arrival and departure curves as

$$D = \frac{c(8.9 - 4 + \lambda g)}{2} + c(4) - \frac{g(\lambda c + 4 - 2)}{2}$$
$$= 147 + 30\lambda g + 240 - 30\lambda g - g = 387 - g$$

$D = 5.78(60) = 387 - g$

$g = 40.2$ s

by inspection of the queue diagram, $\lambda r = 8.9 - 4$ or $\lambda(60-g) = 4.9$

with $g = 40.2$, $\lambda = 0.247$ veh / s or 889.2 veh / h

PROBLEM 6.8

$\lambda = 0.1389$ veh/s; $\mu = 0.5$ veh/s

$\lambda(t) = 0.1389 + 0.000926\,t$

so number of vehicles

$$\int 0.1389 + 0.000926\,t = 0.1389t + 0.000463\,t^2$$

after one cycle

vehicles $= 0.1389(60) + 0.000463(60)^2 = 10$ veh

0 plus the 16 in queue means that 26 will have arrived and 20 (0.4×40) will have departed. So 6 will be in the queue at the start of the second cycle

First cycle delay:

$$= \int_0^{60} 0.1389t + 0.000463\,t^2\,dt + (16)(60) - \frac{(40)(20)}{2}$$

$$= 0.695(3600) + 0.000154(216000) + 960 - 400 = 843.46 \text{ veh-s}$$

for queue dissipation in the second cycle at time t after the start of the cycle,

$$6 + 0.1389(t + 80) + 0.000463(t + 80)^2 = 20 + \mu t = 20 + 0.5t$$

solving quadratic gives t $= 582.6$ s or 37.25 s, 37.25 is the choice so the queue will clear 117.25 s $(80 + 37.25)$ after the beginnining of the effective red of the first cycle

Second cycle delay:

$$= \int_{60}^{117.25} 0.1389t + 0.000463\,t^2\,dt - (4)(57.25) - \frac{(37.25)(18.6520)}{2}$$

$$= 0.695(13747.6) + 0.000154(1611901.7) - 283.46 - 229 - 347.37 = 343.87 \text{ veh-s}$$

total delay is 1187.33 veh-s $(843.46 + 343.87)$

PROBLEM 6.9

$\mu = 3600/3600 = 1$ veh/s

at queue dissipation

$\lambda(60 - 8) = 1\,(60 - 8 - r)$

and $r = 13/\lambda$

so, $52\lambda = 52 - 13/\lambda$

using quadratic formula $\lambda = 0.5$ veh/s or 1800 veh/h

PROBLEM 6.10

$\mu = 1800/3600 = 0.5$ veh/s

by inspection of a queuing diagram, we find that equating vertical distances gives,

$\lambda(80 - g) + 2\lambda g + 2 = 0.5g$

$80\lambda - g\lambda + 2g\lambda + 2 = 0.5g$

and

$\lambda(80 - g) = 5.9$

$g = 80 - 5.9/\lambda$

with two equations and two unknowns we have

$80\lambda - 80\lambda + 5.9 + 160\lambda - 11.8 + 2 = 40 - 2.95/\lambda$

$160\lambda^2 - 43.9\lambda + 2.95 = 0$

solving the quadratic gives $\lambda = 0.118$ or 0.157 veh/s, both of which are feasible. With $\lambda = 0.118$, $g = 30$ s $(80 - 5.9/0.118)$ and at $\lambda = 0.157$, $g = 42.42$ s $(80 - 5.9/0.157)$

with $g = 30$, $r = 50$ and delay is (by inspection of the queuing diagram and using triangles)

$$D = \frac{(2 + 7.9)(50)}{2} + \frac{7.9(30)}{2} = 366 \text{ veh-s}$$

with $g = 42.42$, $r = 37.58$ and delay is (by inspection of the queuing diagram and using triangles)

$$D = \frac{(2 + 7.9)(37.58)}{2} + \frac{7.9(42.42)}{2} = 353.56 \text{ veh-s}$$

PROBLEM 6.11

arrivals $= \int 0.22 + 0.012t = 0.22t + 0.006t^2$

at 60 s arrivals are

$0.22(60) + 0.006(60)^2 = 34.8$

$\lambda c = \mu g$ to clear at the end of the cycle

$\mu = 3600/3600 = 1$ veh/s

$g = 34.8/1 = 34.8$ s

delay is,

$$\int_0^{60} 0.22t + 0.006\,t^2\,dt - \frac{1}{2}(34.8)(34.8) = 0.11(60)^2 + 0.002(60)^3 - 605.52$$

$$222.48 \text{ veh-s}$$

PROBLEM 6.12

$\lambda = 800/3600 = 0.222$ veh/s, first cycle

$\lambda = 500/3600 = 0.139$ veh/s, second cycle

$\mu = 1200/3600 = 0.333$ veh/s

by inspection of a queuing diagram, (with 8.89 vehicles arriving and 6.67 departing at the end of the first cycle and 14.45 and 13.34 arriving and departing at the end of the second cycle) we find that the first cycle delay is (using triangles)

$40(8.89)/2 - 20(6.67)/2 = 111.1$ veh-s

similarly, second cycle delay is,

$2.23(40) + (14.45-8.89)(40)/2 - 20(13.34-6.67)/2 = 133.7$ veh-s

PROBLEM 6.13

For Vine-street phasing

$$YT = t_p + \frac{V}{2a + 2g_r G} = 1.0 + \frac{55 \times 0.2778}{2(3.05) + 2(9.807)(0.08)} = 2.99 \text{ s}$$

all other YT's and AR's are as before. The summation of yellows is now 9 s and the lost time is 15 s (9+6). Using Eq. 6.12

$$c = \frac{1.5LT + 5}{1.0 - \sum_{i=1}^{n} y_i} = \frac{1.5(15) + 5}{1.0 - 0.581} = 65.56 \text{ s}$$

still round up to 70 s so green-time calculations are

Vine-Street phase: (377/1045)(70-15)   = 19.84 s

Vine-Street left: (304/1045)(70-15)   = 16 s

Vine-Street phase: (364/1045)(70-15)   = 19.84 s

this would make:

GT Vine-Street phase   20 s

GT Vine-Street left   16 s

GT Maple-Street phase   19 s

YT Vine-Street phase   3 s

YT Vine-Street left   3 s

All other values as in Table 6.2. Pedestrian times still check out.

PROBLEM 6.14

Equivalent straight-through passenger cars for left turns are:

$200(1.6) + 25(1.6)(1.5) = 380$

Left-turn phase critical-lane volume becomes 380. Yellow and all red times remain the same. Using Eq. 6.12

$$c = \frac{1.5LT + 5}{1.0 - \sum\limits_{i=1}^{n} y_i} = \frac{1.5(16) + 5}{1.0 - \left[\dfrac{377}{1800} + \dfrac{380}{1800} + \dfrac{364}{1800} \right]} = 76.88 \text{ s}$$

round up to 80 s, so green-time calculations are

Vine-Street phase: (377/1121)(64)   = 21.52 s or 21 s

Vine-Street left: (380/1121)(64)   = 21.69 s or 22 s

Vine-Street phase: (364/1121)(64)   = 20.78 s or 21 s

All-red and yellow times do not change. Pedestrian times still check out.

PROBLEM 6.15

Vine-Street critical volumes:

northbound = 699/3 = 233; southbound = 753.6/3 = 251.2 or 252

Maple Street all-red becomes

$$AR = \frac{w + l}{V} = \frac{21.6 + 6}{40 \times 0.2778} = 2.48 \text{ s}$$

Set all-red to 2.5 s, so $LT = 16.5$ s. Other yellow times and all-red times are not changed.

$$c = \frac{1.5LT + 5}{1.0 - \sum_{i=1}^{n} y_i} = \frac{1.5(16.5) + 5}{1.0 - \left[\dfrac{252}{1800} + \dfrac{304}{1800} + \dfrac{364}{1800}\right]} = 60.85 \text{ s}$$

round up to 65 s, so green-time calculations are

Vine-Street phase: (252/920)(48.5)   = 13.29 s or 13.5 s

Vine-Street left: (304/920)(48.5)   = 16.02 s or 16 s

Vine-Street phase: (364/920)(48.5)   = 19.2 s or 19 s

Check pedestrian green time,

$$PGT = 7 + \frac{w}{PWS} - YT - AR$$

for Vine $PGT = 7 + \dfrac{18.3}{1.2} - 3.5 - 2 = 16.75 \text{ s}$

for Maple $PGT = 7 + \dfrac{21.6}{1.2} - 3.5 - 2 = 19.5 \text{ s}$

More green time is needed on Vine and Maple for pedestrians. Increase the cycle to 70 s giving 53.5 s of green time. Let GT Vine phase = 17s, GT Vine left = 16.5 s, and GT Maple phase = 20 s.

PROBLEM 6.16

For all red,

$$AR = \frac{w + l}{V} = \frac{w + 6}{45 \times 0.2778} = 2.0 \text{ s}$$

$$w = 19 \text{ m}$$

for pedestrian green time with the walking speed = 0.9 m/s

$$PGT = 7 + \frac{w}{PWS} - YT - AR$$

$$20 = 7 + \frac{w}{0.9} - 3 - 2$$

$$w = 16.2 \text{ m}$$

so 16.2 m is the widest possible

PROBLEM 6.17

$$c = 60 = \frac{1.5LT + 5}{1.0 - \sum_{i=1}^{n} y_i} = \frac{1.5(16) + 5}{1.0 - \left[\frac{200}{1800} + \frac{187}{1800} + \frac{210}{1800} + \frac{V}{1800} \right]}$$

$$= 332.4 \text{ or } 333$$

$$GT = \frac{333}{930}(44) = 15.75 \text{ s or } 16 \text{ s}$$

SOLUTIONS: CHAPTER 7

PROBLEM 7.1

Using Eq. 7.6

$$SF_i = c_j \times (v/c)_i \times N \times f_w \times f_{HV} \times f_p$$

N	= 3 (given)
c_j	= 2300 pcphpl (six-lane freeway)
v/c	= 0.673 (Table 7.1)
f_w	= 0.94 (Table 7.2)
f_p	= 1.0 (Table 7.7)
E_T	= 3.0 (Table 7.3)

$$f_{HV} = \frac{1}{1 + P_T(E_T - 1)} = \frac{1}{1 + 0.1(3-1)} = 0.833$$

SF = (2300)(0.673)(3)(0.94)(0.833)(1.0) = 3637.54

SF = V/PHF

V = 3637.56(0.90) = 3273.8 veh/h

so ADT = 3273.8/0.20 = 16,369

PROBLEM 7.2

At LOS E v/c = 0.849

SF = 3500/0.9 = 3888.89

Using Eq. 7.6

$$SF_i = c_j \times (v/c)_i \times N \times f_w \times f_{HV} \times f_p$$

N	= 3 (given)
c_j	= 2300 pcphpl (six-lane freeway)
v/c	= 0.849 (Table 7.1)
f_w	= 0.94 (Table 7.2)
f_p	= 1.0 (Table 7.7)

solving for f_{HV}

$$f_{HV} = \frac{SF_i}{c_j \times (v/c)_i \times N \times f_w \times f_p} = \frac{3888.89}{2300 \times 0.849 \times 3 \times 0.94 \times 1.0} = 0.706$$

$$f_{HV} = \frac{1}{1 + P_T(E_T - 1)} = \frac{1}{1 + 0.1(E_T - 1)} = 0.706$$

$E_T = 5.16$

from Table 7.4 E_T at 1/2 - 3/4 = 5.0 and 3/4 - 1 = 5.5. So the highway will drop to LOS E if the grade is 3/4 mi or longer.

PROBLEM 7.3

SF   = 4×700 = 2800

N = 2 (given)

c_j = 2200 pcphpl (four-lane freeway)

v/c   = 0.747 (Table 7.1)

f_w = 1.0 (Table 7.2)

f_p = 1.0 (Table 7.7)

E_T = 3.0 (Table 7.3)

$$f_{HV} = \frac{1}{1 + P_T(E_T - 1)} = \frac{1}{1 + 2P_T}$$

$P_T = N_T/1800$, where N_T is the number of trucks.

Using Eq. 7.6

$$SF_i = c_j \times (v/c)_i \times N \times f_w \times f_{HV} \times f_p$$

$$2800 = 2200 \times 0.747 \times 2 \times 1.0 \times \frac{1}{1 + \dfrac{2N_T}{1800}} \times 1.0$$

$N_T = 156.38$

So 156 trucks and buses can be added.

PROBLEM 7.4

SF   = 4×700 = 2800

N = 2 (given)

c_j = 2200 pcphpl (four-lane freeway)

v/c $= 0.673$ (Table 7.1)

f_w $= 1.0$ (Table 7.2)

f_p $= 1.0$ (Table 7.7)

E_T $= 4.0$ (Table 7.3)

$$f_{HV} = \frac{1}{1 + P_T(E_T - 1)} = \frac{1}{1 + 0.1(4 - 1)} = 0.769$$

Using Eq. 7.6

$$v/c = \frac{SF_i}{c_j \times N \times f_w \times f_{HV} \times f_p} = \frac{2800}{2200 \times 2 \times 1.0 \times 0.769 \times 1.0} = 0.828$$

So LOS D from Table 7.1, because $0.747 < 0.828 < 0.916$.

PROBLEM 7.5

SF $= 2470/0.80 = 3087.5$

N $= 3$ (given)

c_j $= 2300$ pcphpl (six-lane freeway)

v/c $= 1.0$ (Table 7.1)

f_w $= 0.86$ (Table 7.2)

f_p $= 1.0$ (Table 7.7)

E_T $= 7.0$ (Table 7.4)

E_R $= 3.0$ (Table 7.5)

$$f_{HV} = \frac{1}{1 + P_T(E_T - 1) + P_R(E_R - 1)} = \frac{1}{1 + 0.08(7 - 1) + 0.06(3 - 1)} = 0.625$$

Using Eq. 7.6

$$f_p = \frac{SF_i}{c_j \times v/c \times N \times f_w \times f_{HV}} = \frac{3087.5}{2300 \times 1.0 \times 3 \times 0.86 \times 0.625} = 0.832$$

PROBLEM 7.6

N $= 2$ (given)

c_j $= 2200$ pcphpl (four-lane freeway)

v/c $= 1.0$

f_w = 0.93 (Table 7.2) with (5+3)/2 = 4 ft

f_p = 0.9

E_T = 6.0 (Table 7.3)

E_R = 4.0 (Table 7.3)

$$f_{HV} = \frac{1}{1 + P_T(E_T - 1) + P_R(E_R - 1)} = \frac{1}{1 + 0.12(6-1) + 0.06(4-1)} = 0.56$$

Using Eq. 7.6

$$SF_i = c_j \times v/c \times N \times f_w \times f_{HV} \times f = 2200 \times 1.0 \times 2 \times 0.93 \times 0.56 \times 0.9 = 2062.37$$

for the six-lane case $N = 2$ and $c_j = 2300$ and all other factors are unchanged

$$v/c = \frac{SF_i}{c_j \times N \times f_w \times f_{HV} \times f_p} = \frac{2062.37}{2300 \times 3 \times 0.93 \times 0.56 \times 0.90} = 0.638$$

So LOS D from Table 7.1, because 0.626 < 0.638 < 0.793.

PROBLEM 7.7

A distance weighted average of grades can be used because grades are 4 % or less and total length is less than 4000 ft.

$((3 \times 1500) + (4 \times 1000))/2500 = 3.4\%$ for 0.47 mi (2500/5280)

From Table 7.4

E_T at 3% grade, 5% trucks, 1/4-1/2 mi = 2.5

E_T at 4% grade, 5% trucks, 1/4-1/2 mi = 4.0

interpolating, $E_T = 3.1$

$$f_{HV} = \frac{1}{1 + P_T(E_T - 1)} = \frac{1}{1 + 0.05(3.1 - 1)} = 0.905$$

N = 2 (given)

c_j = 2200 pcphpl (four-lane freeway)

f_w = 0.965 (Table 7.2, (0.98+0.95)/2)

f_p = 1.0

SF   = 2000/0.90 = 2222.22

$$v/c = \frac{SF_i}{c_j \times N \times f_w \times f_{HV} \times f_p} = \frac{2222.22}{2200 \times 2 \times 0.965 \times 0.905 \times 1.0} = 0.578$$

So LOS C from Table 7.1, because 0.436 < 0.578 < 0.655.

PROBLEM 7.8

Using initial conditions there are 2200(0.15) = 330 trucks and buses in the traffic stream. This gives a truck and bus SF of 419.85 (330/0.786). The equation for f_{HV} is

$$f_{HV} = \frac{1}{1 + P_T(E_T - 1)} = \frac{1}{1 + \dfrac{419.85}{SF}(6 - 1)}$$

Using Eq. 7.6 with factors taken from Example 7.2

$$SF = 2300 \times 1.0 \times 3 \times 0.86 \times \left(\frac{1}{1 + \dfrac{419.85}{SF}(6 - 1)} \right) \times 1.0 = 3834.77$$

$V = 3834.77 \times 0.786 = 3014.13$ veh/h or 814 passenger cars (3014.13 - 2200).

PROBLEM 7.9

with trucks SF = 5400/0.95 = 5684.21

N = 4 (given)

c_j = 2300 pcphpl (eight-lane freeway)

f_w = 0.93 (Table 7.2)

f_p = 1.0

E_T = 3.0 (Table 7.3)

$$f_{HV} = \frac{1}{1 + P_T(E_T - 1)} = \frac{1}{1 + 0.11(3 - 1)} = 0.82$$

Using Eq. 7.6

$$v/c = \frac{SF_i}{c_j \times N \times f_w \times f_{HV} \times f_p} = \frac{5684.21}{2300 \times 4 \times 0.93 \times 0.82 \times 1.0} = 0.81$$

So LOS E from Table 7.1, because 0.793 < 0.81 < 1.0.

eliminating trucks gives $V = 5400 - 0.6(5400) = 5076$. Percent buses = (0.5×5400/5076) = 0.053.

$$f_{HV} = \frac{1}{1 + P_T(E_T - 1)} = \frac{1}{1 + 0.053(3 - 1)} = 0.904$$

SF = 5076/0.95 = 5343.16

$$v/c = \frac{SF_i}{c_j \times N \times f_w \times f_{HV} \times f_p} = \frac{5343.16}{2300 \times 4 \times 0.93 \times 0.904 \times 1.0} = 0.69$$

So LOS D from Table 7.1, because 0.626 < 0.691 < 0.793.

PROBLEM 7.10

Before:

SF = 3800/0.9 = 4222.22

N	= 3 (given)
c_j	= 2300 pcphpl (eight-lane freeway)
f_w	= 1.0 (Table 7.2)
f_p	= 1.0 (Table 7.7)
E_T	= 8.0 (Table 7.4)

$$f_{HV} = \frac{1}{1 + P_T(E_T - 1)} = \frac{1}{1 + 0.06(8 - 1)} = 0.704$$

Using Eq. 7.6

$$v/c = \frac{SF_i}{c_j \times N \times f_w \times f_{HV} \times f_p} = \frac{4222.22}{2300 \times 3 \times 1.0 \times 0.704 \times 1.0} = 0.869$$

So LOS E from Table 7.1, because 0.849 < 0.869 < 1.0.

eliminating buses

Number of buses = 0.04×3800 = 152

Number of cars added = 6×152 = 912, so new V = 3800 + 912 - 152 = 4560.

New P_T =0.0166 ((0.02×3800)/4560) and new E_T = 13.0 (Table 7.4, rounding up to 2 %)

$$f_{HV} = \frac{1}{1 + P_T(E_T - 1)} = \frac{1}{1 + 0.0166(13 - 1)} = 0.834$$

SF = 4560/0.9 = 5066.67

$$v/c = \frac{SF_i}{c_j \times N \times f_w \times f_{HV} \times f_p} = \frac{5066.67}{2300 \times 3 \times 1.0 \times 0.834 \times 1.0} = 0.880$$

So LOS E from Table 7.1, because 0.849 < 0.880 < 1.0.

PROBLEM 7.11

PHF = 0.85

E_T = 1.5 (Table 7.3)

E_R = 1.2 (Table 7.3)

$$f_{HV} = \frac{1}{1 + P_T(E_T - 1) + P_R(E_R - 1)} = \frac{1}{1 + 0.06(1.5 - 1) + 0.02(1.2 - 1)} = 0.967$$

from Eq. 7.11

$$v_p = \frac{V}{(N)(\text{PHF})(f_{HV})} = \frac{1300}{(2)(0.85)(0.967)} = 790.8$$

So LOS B from Table 7.8, because 540 < 790.8 < 900.

PROBLEM 7.12

from Table 7.8 v_p = 1900 at capacity

E_T = 1.5 (Table 7.3)

E_R = 1.2 (Table 7.3)

$$f_{HV} = \frac{1}{1 + P_T(E_T - 1) + P_R(E_R - 1)} = \frac{1}{1 + 0.06(1.5 - 1) + 0.02(1.2 - 1)} = 0.967$$

from Eq. 7.11

$$v_p = \frac{V}{(N)(\text{PHF})(f_{HV})} = \frac{V}{(2)(0.85)(0.967)} = 1900$$

$V = 3123$

So 1823 vehicles (3123-1300).

PROBLEM 7.13

FFS = 49.9

F_M = 1.6 (Table 7.10)

F_{LW} = 6.6 (Table 7.11)

TLC = 2 + 6 = 8 (Eq. 7.9)

F_{LC} = 0.9 (Table 7.12)

F_A $= 0.25 \times 4 = 1.0$ (Eq. 7.10)

using Eq. 7.8,

$$FFS = FFS_I - F_M - F_{LW} - F_{LC} - F_A$$

$FFS_I = 49.9 + 1.6 + 6.6 + 0.9 + 1.0 = 60$ mph

from Table 7.9, assume speed limit 50 or 55, speed limit = 60 - 5 = 55 mph

85th percentile = 60 -[3 - 0.1(85th percentile)] = 57/1.1 = 51.82 mph

PROBLEM 7.14

$$FFS = FFS_I - F_M - F_{LW} - F_{LC} - F_A$$

from Table 7.9 $FFS_I = 5 + 55 = 60$

F_M $= 0.0$ (Table 7.10)

F_{LW} $= 1.9$ (Table 7.11)

TLC $= 4 + 6 = 10$ (Eq. 7.9)

F_{LC} $= 0.4$ (Table 7.12)

F_A $= 0.25 \times 13 = 3.25$ (Eq. 7.10)

using Eq. 7.8,

$FFS = 60 - 0 - 1.9 - 0.4 - 3.25 = 54.45$ mph

E_T $= 6.0$ (Table 7.4)

E_R $= 3.0$ (Table 7.5)

$$f_{HV} = \frac{1}{1 + P_T(E_T - 1) + P_R(E_R - 1)} = \frac{1}{1 + 0.08(6 - 1) + 0.02(3 - 1)} = 0.694$$

PHF = 0.90 (given)

$$v_p = \frac{V}{(N)(\text{PHF})(f_{HV})} = \frac{3900}{(3)(0.90)(0.694)} = 2081.33$$

from Fig. 7.5 LOS E

PROBLEM 7.15

before development:

FFS_I $= 58 + [3 - 0.1(58)] = 55.2$ (Table 7.9)

F_M $= 0.0$ (Table 7.10)

59

F_{LW} = 6.6 (Table 7.11)

TLC = 6 + 3 = 9 (Eq. 7.9)

F_{LC} = 0.65 (Table 7.12, (0.9+0.4)/2)

F_A = 0.25×4 = 1.0 (Eq. 7.10)

using Eq. 7.8,

FFS = 55.2 - 0 - 6.6 - 0.65 - 1 = 46.95 mph

E_T = 3.0 (Table 7.3)

E_R = 2.0 (Table 7.3)

$$f_{HV} = \frac{1}{1 + P_T(E_T - 1) + P_R(E_R - 1)} = \frac{1}{1 + 0.10(3-1) + 0.03(2-1)} = 0.813$$

PHF = 0.90 (given)

$$v_p = \frac{V}{(N)(PHF)(f_{HV})} = \frac{2300}{(2)(0.95)(0.813)} = 1488.96$$

from Fig. 7.5 LOS D.

after development

FFS = 46.95 - (0.225×8) = 44.95 mph

$$v_p = \frac{V}{(N)(PHF)(f_{HV})} = \frac{2700}{(2)(0.95)(0.813)} = 1747.91$$

from Fig. 7.5 LOS E

PROBLEM 7.16

before E_T = 12.5 (Table 7.4)

$$f_{HV} = \frac{1}{1 + P_T(E_T - 1)} = \frac{1}{1 + 0.02(12.5 - 1)} = 0.813$$

$$v_p = \frac{V}{(N)(PHF)(f_{HV})} = \frac{1900}{(2)(0.8)(0.813)} = 1460.6$$

from Table 7.8 LOS C (1100 < 1460.6 < 1510)

with trucks, number of buses = 1900×0.02 = 38

new P_T = (38+150)/(1900+150) = 0.092

E_T = 7.0 (Table 7.4)

$$f_{HV} = \frac{1}{1+P_T(E_T-1)} = \frac{1}{1+0.092(7-1)} = 0.644$$

$$v_p = \frac{V}{(N)(PHF)(f_{HV})} = \frac{2050}{(2)(0.8)(0.644)} = 1989.5$$

from Table 7.8 LOS E (1800 < 1989.5 < 2100)

PROBLEM 7.17

FFS_I = 5 + 55 = 60

F_M = 1.6 (Table 7.10)

F_{LW} = 1.9 (Table 7.11)

TLC = 6 + 4 = 10 (Eq. 7.9)

F_{LC} = 0.4 (Table 7.12)

F_A = 0.25×10 = 2.5 (Eq. 7.10)

using Eq. 7.8,

FFS = 60 - 1.6 - 1.9 - 0.4 - 2.5 = 53.6 mph

E_T = 1.5 (Table 7.3)

$$f_{HV} = \frac{1}{1+P_T(E_T-1)} = \frac{1}{1+0.08(1.5-1)} = 0.962$$

interpolate between 50 and 55 to get v_p = 2072

$$v_p = \frac{V}{(N)(PHF)(f_{HV})} = \frac{V}{(2)(0.8)(0.962)} = 2072$$

= 3189.22 veh

PROBLEM 7.18

FFS_I = 5 + 50 = 55

F_M = 0.0 (Table 7.10)

F_{LW} = 0.0 (Table 7.11)

F_{LC} = 0.0 (Table 7.12)

F_A = 0.25×4 = 2 (Eq. 7.10)

using Eq. 7.8,

$FFS = 55 - 2 = 53$ mph

$V = 1800 + 140 + 40 + 10 = 1990$

$P_T = (140+40)/1990 = 0.09; P_R = (10)/1990 = 0.005$

$E_T \qquad = 1.5$ (Table 7.6)

$E_R \qquad = 1.2$ (Table 7.3)

$$f_{HV} = \frac{1}{1 + P_T(E_T - 1) + P_R(E_R - 1)} = \frac{1}{1 + 0.09(1.5 - 1) + 0.005(1.2 - 1)} = 0.956$$

PHF = 0.85 (given)

$$v_p = \frac{V}{(N)(PHF)(f_{HV})} = \frac{1990}{(2)(0.85)(0.956)} = 1224.46$$

LOS C from Fig. 7.5

PROBLEM 7.19

$E_T \qquad = 6.0$ (Table 7.3)

$E_R \qquad = 4.0$ (Table 7.3)

$$f_{HV} = \frac{1}{1 + P_T(E_T - 1) + P_R(E_R - 1)} = \frac{1}{1 + 0.06(6 - 1) + 0.02(4 - 1)} = 0.735$$

at maximum LOS C $v_p = 1400$ (Table 7.8), so

$\qquad = v_p(N)(PHF)(f_{HV}) = 1400(3)(0.9)(0.735) = 2778.3$

at capacity with PHF = 0.95, $v_p = 2000$ (Table 7.8)

$\qquad = v_p(N)(PHF)(f_{HV}) = 2000(3)(0.95)(0.735) = 4189.5$

Number added is 4189.5 - 2778.3 = 1411.2, about 1411 vehicles

PROBLEM 7.20

$NAPM = 13/0.59 = 22$

$FFS_I \quad = 60 + [3 - 0.1(60)] = 57$

$F_M \quad = 1.6$ (Table 7.10)

$F_{LW} \quad = 1.9$ (Table 7.11)

$TLC \quad = 6 + 5 = 11$ (Eq. 7.9)

F_{LC} = 0.2 (Table 7.12)

F_A = 0.25×22 = 5.5 (Eq. 7.10)

using Eq. 7.8,

FFS = 57 - 1.6 - 1.9 - 0.2 - 5.5 = 47.8 mph

PHF = 2212/2400 = 0.922

P_T = (212)/2212 = 0.0958

E_T = 5.105 by interpolating (Table 7.4)

$$f_{HV} = \frac{1}{1 + P_T(E_T - 1)} = \frac{1}{1 + 0.0958(5.105 - 1)} = 0.718$$

$$v_p = \frac{V}{(N)(\text{PHF})(f_{HV})} = \frac{2212}{(2)(0.922)(0.718)} = 1670.7$$

from Fig. 7.5 LOS D is reached at FFS about 50 mph. So 50 - 47.8 = 2.2, this gives 8.8 access points per mile (2.2/0.25). The number of access points must be reduced from 22/mi to 13/mi. or 13×0.59 = 7.67 or 7 to be conservative so 6 access points must be eliminated.

PROBLEM 7.21

SF = 256×4 = 1024

v/c = 0.78 (100 percent no-passing mountainous, Table 7.15)

f_w = 0.88 (Table 7.16)

f_d = 0.89 (Table 7.14)

using Eq. 7.12

$$f_{HV} = \frac{SF_i}{2800 \times (v/c)_i \times f_d \times f_w} = \frac{1024}{2800 \times 0.78 \times 0.89 \times 0.88} = 0.60$$

E_T = 12 (Table 7.17)

$$f_{HV} = \frac{1}{1 + P_T(E_T - 1)} = \frac{1}{1 + P_T(12 - 1)} = 0.60$$

P_T = 0.061

PROBLEM 7.22

SF = 180/0.87 = 207

must solve

$$v/c = \frac{SF_i}{2800 \times f_d \times f_w \times f_{HV}}$$

LOS	f_d	f_w	E_T	E_R	f_{HV}	v/c
A	0.94	0.75	7	5	0.588	0.178
B	0.94	0.75	10	5.2	0.535	0.196
C	0.94	0.75	10	5.2	0.535	0.196
D	0.94	0.75	12	5.2	0.508	0.206
E	0.94	0.88	12	5.2	0.508	0.176

f_{HV} computed with $P_T = 0.05$ and $P_R = 0.10$.

So LOS C from Table 7.15 is a match $0.12 < 0.196 < 0.20$ and so is LOS D.

PROBLEM 7.23

SF $= 280/0.8 = 350$

v/c $= 0.35$ (Table 7.15)

f_w $= 0.68$ (Table 7.16)

$E_T = 5.0$ (Table 7.17), $E_B = 3.9$ (Table 7.17), $E_R = 3.4$ (Table 7.17)

using Eq. 7.12

$$f_{HV} = \frac{1}{1 + P_T(E_T - 1) + P_B(E_B - 1) + P_R(E_R - 1)}$$

$$= \frac{1}{1 + 0.1(5 - 1) + 0.02(3.9 - 1) + 0.05(3.4 - 1)}$$

$$= 0.634$$

$$f_d = \frac{SF_i}{2800 \times (v/c)_i \times f_{HV} \times f_w} = \frac{350}{2800 \times 0.35 \times 0.634 \times 0.68} = 0.828$$

from Table 7.14 the worst allowable directional split is 80/20 for LOS C

PROBLEM 7.24

directional split 480/800 so 60/40

SF $\quad = 800/0.75 = 1066.67$

$f_d \quad = 0.94$ (Table 7.14)

$f_w \quad = 0.85$ (Table 7.16, for LOS A-D)

$f_w \quad = 0.92$ (Table 7.16, for LOS E)

at LOS A

$$f_{HV} = \frac{1}{1+0.04(6)+0.01(4)+0.03(4.7)} = 0.704$$

at LOS B-C

$$f_{HV} = \frac{1}{1+0.04(9)+0.01(4.2)+0.03(5)} = 0.644$$

at LOS D-E

$$f_{HV} = \frac{1}{1+0.04(11)+0.01(4.2)+0.03(5.5)} = 0.607$$

assuming LOS A

$$v/c = \frac{1066.67}{2800 \times 0.85 \times 0.94 \times 0.704} = 0.677 \text{ or LOS E}$$

assuming LOS B-C

$$v/c = \frac{1066.67}{2800 \times 0.85 \times 0.94 \times 0.644} = 0.738 \text{ or LOS E}$$

assuming LOS A

$$v/c = \frac{1066.67}{2800 \times 0.92 \times 0.94 \times 0.607} = 0.726 \text{ or LOS E}$$

so is LOS E

PROBLEM 7.25

$f_p = f_w = f_{HV} = 1.0$ all ideal conditions

From Fig. 7.9, $K_{10} = 0.135$, $K_{50} = 0.112$, $K_{100} = 0.106$

from Eq. 7.14

$DDHV_{10} = 0.135 \times 0.7 \times 30000 = 2835$

$DDHV_{50} = 0.112 \times 0.7 \times 30000 = 2352$

$DDHV_{100} = 0.106 \times 0.7 \times 30000 = 2226$

$SF_{10} = 2835/0.8 = 3543.75$

$SF_{50} = 2352/0.8 = 2940$

$SF_{100} = 2226/0.8 = 2782.5$

$v/c_{10} = 3543.75/(2200 \times 2) = 0.805$

$v/c_{50} = 2940/(2200 \times 2) = 0.668$

$v/c_{100} = 2782.5/(2200 \times 2) = 0.632$

from Table 7.1, 10th highest is LOS D, 50th highest is LOS C, 100th highest is LOS C

PROBLEM 7.26

N = 2 (given)

c_j = 2200 pcphpl (four-lane freeway)

v/c   = 1.0 (Table 7.1)

f_w = 0.93 (Table 7.2)

f_p = 1.0 (Table 7.7)

E_T = 3.0 (Table 7.3)

$$f_{HV} = \frac{1}{1 + P_T(E_T - 1)} = \frac{1}{1 + 0.08(3-1)} = 0.862$$

$SF = (2300)(1.0)(2)(0.93)(0.862)(1.0) = 3527.3$

$V = SF \times PHF = 3527.3 \times 0.85 = 2998.2$ veh/h

so $AADT = 2998.2/(K \times D) = 2998.2/(0.12 \times 0.6) = 41642$

SOLUTIONS: CHAPTER 8

PROBLEM 8.1

Trips per household:

shopping   = 0.12 + 0.09(2) + 0.011(20) - 15(1) = 0.37

social/rec   = 0.04 + 0.018(2) + 0.009(20) + 0.16 (2) = 0.576

So, shopping trips = 624 (1700×0.37) and social/recreation trips = 979.2 (1700×0.576). Total trips = 1680.2

PROBLEM 8.2

Shopping trips per household × 1700 households = 100

So, $(0.012 + 0.09(2) + 0.011(20) - 0.15x) \times 1700 = 100$

$x = 3.07$, which means 207 additional employees (307-100)

PROBLEM 8.3

Base number of trips = 0.04 + 0.018(5.2) + 0.009(15) + 0.16 (4) = 0.9086

Number of new trips = 0.04 + 0.018(5.2) + 0.009(15×1.1) + 0.16 (5.2 -1.2(1.2)) = 0.8837

So, 0.9086(1500) - 0.8837(1500) = - 37.35

PROBLEM 8.4

Vehicles   =   Drive alone = 2380

                   Shared ride = 870/2 = 435

                   Bus = 750/15 = 50

Total number of vehicles = 2870

PROBLEM 8.5

$U_{DL} = 2.2 - 0.02(8) - 0.03(20) = 0$ $\qquad e^U = 1$

$U_{SR} = 0.8 - 0.2(4) - 0.03(20) = -0.6$ $\qquad e^U = 0.549$

$U_B = -0.35$ as before $\qquad e^U = 0.705$

$$P_{DL} = \frac{1}{1 + 0.549 + 0.705} = 0.444 \text{ which gives } 1775 \ (0.444 \times 4000)$$

$$P_{SR} = \frac{0.549}{1 + 0.549 + 0.705} = 0.244 \text{ which gives } 974 \ (0.244 \times 4000)$$

$$P_B = \frac{0.705}{1 + 0.549 + 0.705} = 0.705 \text{ which gives } 1251 \ (0.705 \times 4000)$$

PROBLEM 8.6

$U_1 = -0.283(4) + 0.172(20) = 2.308$ $\qquad e^U = 10.05$

$U_2 = -0.283(9) + 0.172(15) = 0.033$ $\qquad e^U = 1.034$

$U_3 = -0.283(8) + 0.172(30) = 2.896$ $\qquad e^U = 18.102$

$U_4 = -0.283(14) + 0.172(60) = 6.358$ $\qquad e^U = 577.09$

$$P_1 = \frac{10.05}{606.28} = 0.0166 \text{ which gives } 66.4 \text{ trips } (0.0166 \times 4000)$$

$$P_2 = \frac{1.034}{606.28} = 0.0017 \text{ which gives } 6.8 \text{ trips } (0.0017 \times 4000)$$

$$P_3 = \frac{18.102}{606.28} = 0.0299 \text{ which gives } 119.6 \text{ trips } (0.0299 \times 4000)$$

$$P_4 = \frac{577.09}{606.28} = 0.9518 \text{ which gives } 3807.2 \text{ trips } (0.9518 \times 4000)$$

PROBLEM 8.7

$U_1 = -0.283(4) + 0.172(20) = 2.308$ $e^U = 10.05$

$U_2 = -0.283(9) + 0.172(50) = 6.053$ $e^U = 425.387$

$U_4 = -0.283(14) + 0.172(60) = 6.358$ $e^U = 577.09$

$P_1 = \dfrac{10.05}{1012.53} = 0.0099$ which gives 39.6 trips (0.0099×4000)

$P_2 = \dfrac{425.387}{1012.53} = 0.4201$ which gives 1680.4 trips (0.4201×4000)

$P_4 = \dfrac{577.09}{1012.53} = 0.5700$ which gives 2280 trips (0.5700×4000)

PROBLEM 8.8

$U_4 = -0.283(14) + 0.172(60) = 6.358$ $e^U = 577.09$

So,

$U_1 = 6.358 = -0.283(4) + 0.172(x)$

$x = 43,546.5$ m² must be added to shopping center 1

$U_2 = 6.358 = -0.283(9) + 0.172(x)$

$x = 51,773.3$ m² must be added to shopping center 1

PROBLEM 8.9

A 20 percent reduction in travel time gives 12 min by auto and 17.6 min by bus

U_{auto} to shopping center # 1 = 1.2 as before $e^U = 3.32$

U_{trans} to shopping center # 1 = -1.2 as before $e^U = 0.301$

U_{auto} to shopping center # 2 = 0.6 - 0.3(12) + 0.12 (40) = 1.8 $e^U = 6.05$

U_{trans} to shopping center # 2 = - 0.3(17.6) + 0.12 (40) = -0.48 $e^U = 0.619$

P_{auto} to shopping center # 1 = $\dfrac{3.32}{10.289}$ = 0.323 which gives 290 trips (0.323×900)

P_{transit} to shopping center # 1 = $\dfrac{0.301}{10.289}$ = 0.0293 which gives 26 trips (0.0293×900)

P_{auto} to shopping center # 2 = $\dfrac{6.05}{10.289}$ = 0.588 which gives 529 trips (0.588×900)

P_{transit} to shopping center # 2 = $\dfrac{0.6188}{10.289}$ = 0.06 which gives 54 trips (0.06×900)

PROBLEM 8.10

Before construction $P_{2A} + P_{2T} = 0.394 + 0.026 = 0.42$

For shopping destination e^U's are the same as before $3.32 + 0.301 = 3.621$

$U_{2A} = 0.6 - 0.3(19) + 0.12\,(x)$

$U_{2T} = -0.3(26) + 0.12\,(x)$

$$0.42 = \frac{e^{U_{2A}}}{3.621 + e^{U_{2A}} + e^{U_{2T}}} + \frac{e^{U_{2T}}}{3.621 + e^{U_{2A}} + e^{U_{2T}}}$$

$$\ln\left[e^{U_{2A}} + e^{U_{2T}}\right] = \ln\left[152 / 0.58\right] = 0.9634$$

$$\ln\left[e^{0.6-0.3(19)+0.12x} + e^{-0.3(26)+0.12x}\right] = 0.9634$$

$x = 49.978$

so 9,978.3 m² (49,978 - 40,000) must be added

PROBLEM 8.11

$U_{\text{DEST 1}} = 0.2(15.5) - 0.15(12) + 0.9(0.5) = 1.75$ $e^U = 5.75$

$U_{\text{DEST 2}} = 0.2(6.0) - 0.15(8) + 0.9(1.0) = 0.9$ $e^U = 2.46$

$U_{\text{DEST 3}} = 0.2(0.8) - 0.15(3) + 0.9(0.8) = 0.43$ $e^U = 1.54$

$U_{\text{DEST 4}} = 0.2(5) - 0.15(11) + 0.9(1.5) = 0.7$ $e^U = 2.01$

$$P_{\text{DEST 1}} = \frac{5.75}{11.76} = 0.489 \text{ which gives } 342 \text{ trips } (0.489 \times 700)$$

$$P_{\text{DEST 2}} = \frac{2.46}{11.76} = 0.209 \text{ which gives } 146 \text{ trips } (0.209 \times 700)$$

$$P_{\text{DEST 3}} = \frac{1.54}{11.76} = 0.131 \text{ which gives } 92 \text{ trips } (0.131 \times 700)$$

$$P_{\text{DEST 4}} = \frac{2.01}{11.76} = 0.171 \text{ which gives } 120 \text{ trips } (0.171 \times 700)$$

PROBLEM 8.12

$U_{\text{DEST 1}} = 0.2(15.5) - 0.15(12) + 0.9(0.5) = 1.75 \qquad e^U = 5.75$

$U_{\text{DEST 2}} = 0.2(6.0) - 0.15(8) + 0.9(1.0) = 0.9 \qquad e^U = 2.46$

$U_{\text{DEST 3}} = 0.2(0.8) - 0.15(3) + 0.9(2.3) = 1.78 \qquad e^U = 5.93$

$U_{\text{DEST 4}} = 0.2(5) - 0.15(11) + 0.9(1.5) = 0.7 \qquad e^U = 2.01$

$$P_{\text{DEST 1}} = \frac{5.75}{16.05} = 0.356 \text{ which gives } 249 \text{ trips } (0.356 \times 700)$$

$$P_{\text{DEST 2}} = \frac{2.46}{16.05} = 0.152 \text{ which gives } 107 \text{ trips } (0.152 \times 700)$$

$$P_{\text{DEST 3}} = \frac{5.93}{16.05} = 0.367 \text{ which gives } 257 \text{ trips } (0.367 \times 700)$$

$$P_{\text{DEST 4}} = \frac{2.01}{16.05} = 0.241 \text{ which gives } 87 \text{ trips } (0.241 \times 700)$$

PROBLEM 8.13

Let subscript 1A denote destination 1 by auto, and so on.

For 40 percent of trips $P_{1A} + P_{1T} = 0.4$

Utilities for all other destination (i.e., 2 and 3) are the same as before so summation of their e^U's $= 8.365 + 1.76 + 1.195 + 0.13 = 11.45$

$U_{1A} = 0.9 - 0.22(14) + 0.16(12.4) + 1.1\,(x)$

$U_{1T} = -0.22(17) + 0.16(12.4) + 1.1\,(x)$

$$0.4 = \frac{e^{U_{1A}}}{3.621 + e^{U_{1A}} + e^{U_{1T}}} + \frac{e^{U_{1T}}}{3.621 + e^{U_{1A}} + e^{U_{1T}}}$$

$$\ln\left[e^{U_{1A}} + e^{U_{1T}}\right] = \ln\left[7.633\right] = 2.0325$$

$$\ln\left[e^{0.9-0.22(14)+0.16(12.4)+1.1x} + e^{-0.22(17)+0.16(12.4)+1.1x}\right] = 2.0325$$

$x = 1.8523$

so 552.3 m^2 (1,852.3 - 1,300) must be added

PROBLEM 8.14

By definition,

$$P_{DEST\,2} = \frac{e^{U_2}}{e^{U_1} + e^{U_2} + e^{U_3} + e^{U_4}}$$

but e^{U_1}, e^{U_3}, e^{U_4} are the same as before, so

$$P_{DEST\,2} = \frac{e^{U_2}}{9.3 + e^{U_2}} = 250/700 = 0.357$$

if x is the total floor space in thousands of square meters, then

$$_2 = 0.2(6) - 0.15(8) + 0.9x = 0.9\,x$$

$$P_{DEST\,2} = \frac{e^{0.9\,x}}{9.3 + e^{0.9\,x}} = 0.357$$

$x = 1.824$

so 824 m^2 (1,824 - 1,000) must be added

PROBLEM 8.15

Part a:

$x_1 = 9.78$

so, $t_1 = t_2 = 4 + 3(9.78) = 33.34$

$x_2 = 15 - 9.78 = 5.22$

with $33.34 = b + 6(5.22)$, $b = 2.0$

Part b:

when $x_1 = 7$ and $x_2 = 0$, $t_1 = 28$ and $t_2 = 2$

when $x_1 = 0$ and $x_2 = 7$, $t_1 = 4$ and $t_2 = 44$

so both routes might be used

At UE, $t_1 = t_2$

$4 + 3x_1 = 2 + 6x_2$ and $x_2 = 7 - x_1$

$4 + 3x_1 = 2 + 6(7 - x_1)$

$x_1 = 4.44$

so, $x_2 = 7 - x_1 = 2.56$

$t_1 = 4 + 3(4.44) = 17.32$

$t_2 = 2 + 6(2.56) = 17.36$

PROBLEM 8.16

when $x_1 = 4$ and $x_2 = 0$, $t_1 = 12$ and $t_2 = 1$

when $x_1 = 0$ and $x_2 = 4$, $t_1 = 8$ and $t_2 = 9$

so both routes might be used

At UE, $t_1 = t_2$

$8 + x_1 = 1 + 2x_2$ and $x_2 = 4 - x_1$

$8 + x_1 = 1 + 2(7 - x_1)$

$x_1 = 0.33$

so, $x_2 = 4 - x_1 = 3.67$

$t_1 = 8 + 0.33 = 8.33$ or 2,750 veh-min (8.33×330)

$t_2 = 1 + 2(3.67) = 8.33$ or 30,570 veh-min (8.33×3670)

73

Total is 33,320 veh-min

At SO,

minimize $z = (8 + x_1)x_1 + (1 + 2x_2)x_2$ with $x_2 = 4 - x_1$

$= (8 + x_1)x_1 + (1 + 2(4 - x_1))(4 - x_1)$

$= 3x_1^2 - 9x_1 + 36$

$dz/dx_1 = 6x_1 - 9 = 0$

So, $x_1 = 1.5$ and $x_2 = 2.5$

$t_1 = 8 + 1.5 = 9.5$ or 14,250 veh-min (9.5×1500)

$t_2 = 1 + 2(2.5) = 6$ or 15,000 veh-min (6×2500)

Total is 29,250 veh-min

PROBLEM 8.17

Total vehicle hours for UE = 875.97 from solution to Example 8.11.

For SO,

minimize $z = (4 + 1.136x_1)x_1 + (3 + 3.182x_2)x_2$ with $x_1 = 6 - x_2$

$= (4 + 1.136(6 - x_2))(6 - x_2) + (3 + 3.182x_2)x_2$

$= 4.318x_2^2 - 14.632x_2 + 51.264$

$dz/dx_2 = 8.636x_2 - 14.632 = 0$

So, $x_2 = 1.694$ and $x_1 = 6 - 1.694 = 4.306$

$t_1 = (4 + 1.136(4.306)) \times 4306/60 = 638.12$

$t_2 = (3 + 3.182(1.694)) \times 1694/60 = 236.89$

which gives a total of 875.01, so the savings is 0.96 veh-hrs (875.97 - 875.01)

PROBLEM 8.18

Before reconstruction,

minimize $z = (2 + 1.2x_1)x_1 + (4 + 0.5x_2)x_2$ with $x_2 = 3.5 - x_1$

$= (2 + 1.2x_1)x_1 + (4 + 0.5(3.5 - x_1))(3.5 - x_1)$

$= 1.7x_1^2 - 5.5x_1 + 20.125$

$dz/dx_1 = 3.4x_1 - 5.5 = 0$

So, $x_1 = 1.618$ and $x_2 = 3.5 - 1.618 = 1.882$

Total travel time $= 1.618(2 + 1.2(1.168)) + 1.882(4 + 5(1.882)) = 15.6739$ or $15,673.9$ veh-min

After reconstruction:

minimize $z = (2 + 1.2x_1)x_1 + (4 + 1x_2)x_2$ with $x_2 = q - x_1$

$= (2 + 1.2x_1)x_1 + (4 + 1(q - x_1))(q - x_1)$

$= 2.2x_1^2 - 2x_1 + q^2 - 4q - 2qx_1$

$dz/dx_1 = 4.4x_1 - 2 - 2q = 0$

$q = 2.2x_1 - 1$

Also, from travel times:

$(2 + 1.2x_1q)x_1 + (4 + 1x_2)x_2 = 15.6739$ with $x_2 = q - x_1$

$2x_1 + 1.2x_1^2 + 4(q - x_1) + (q - x_1)^2 = 15.6739$

$2.2x_1^2 - 2x_1 + 4 + q^2 - 2qx_1 = 15.6739$

substituting $q = 2.2x_1 - 1$

$2.2x_1^2 - 2x_1 + 4 + (2.2x_1 - 1)^2 - 2(2.2x_1 - 1)x_1 = 15.6739$

$x_1 = 1.954$ by quadratic equation

so, $q = 2.2(1.954) - 1 = 3.298$ and $x_2 = 3.298 - 1.954 = 1.344$

PROBLEM 8.19

$P(h < 5) = 1 - P(h \geq 5)$

$P(h \geq 5) = 0.6 = e^{-x_1(0.5)/3600}$

$\ln(0.6) = -0.00139x_1$

$x_1 = 367.5$ veh/h

so, $t_1 = 2 + 0.3675 = 2.3675$

at UE $t_1 = t_2 = 2.3675$ min, thus $x_2 = 1.3675$ ($2.3675 - 1$, from performance function) or 1367.5 veh/h

75

when $x_1 = 3$ and $x_2 = 0$ and $x_3 = 0$, $t_1 = 9.5$, $t_2 = 1$, $t_3 = 3$

when $x_1 = 0$ and $x_2 = 3$ and $x_3 = 0$, $t_1 = 8$, $t_2 = 7$, $t_3 = 3$

when $x_1 = 0$ and $x_2 = 0$ and $x_3 = 3$, $t_1 = 8$, $t_2 = 1$, $t_3 = 2.25$

So route 1 will never be use. Thus $x_2 + x_3 = 3.0$ and at UE, $t_2 = t_3$

$1 + 2x_2 = 3 + 0.75x_3$ and $x_3 = 3 - x_2$

$1 + 2x_2 = 3 + 0.75(3 - x_2)$

$x_2 = 1.545$

so, $x_3 = 3 - 1.545 = 1.455$

PROBLEM 8.21

Let x_{11} denote route one before construction

Let x_{12} denote route one during construction

Let x_{21} denote route two before construction

Let x_{22} denote route two during construction

we know $t_{12} - t_{11} = 35.28$ s or 0.588 min

Also, because in UE when $t_1 = t_2$; $t_{22} - t_{21} = 0.588$ min

$t_{22} = 10 + 1.5x_{22}$ and $t_{21} = 10 + 1.5x_{21}$

substituting:

$(10 + 1.5x_{22}) - (10 + 1.5x_{21}) = 0.588$

$x_{21} = x_{22} - 0.392$

Also, for traffic increase: $x_{22} = 1.685x_{21}$ or $x_{21} = 0.5935x_{22}$

substituting:

$0.5935x_{22} = x_{22} - 0.392$

$x_{22} = 0.964$

So, $t_{22} = (10 + 1.5(0.964)) = 11.45 = t_{12}$

$t_{12} = 3 + 8(x_{12}/3) = 11.45$ so $x_{12} = 2.044$

also since $t_{12} - t_{11} = t_{22} - t_{21} = 0.588$ min

from UE $t_{11} = t_{21} = 11.45 - 0.588 = 10.86$ min

$t_{11} = 6 + (x_{11}/3) = 10.86$

$x_{11} = 2.43$

$t_{21} = 10 + 3(x_{21}/2) = 10.86$

$x_{21} = 0.573$

PROBLEM 8.22

Part a (with $q = 10000$):

when $x_1 = 10$ and $x_2 = 0$ and $x_3 = 0$, $t_1 = 7$, $t_2 = 1$, $t_3 = 4$

when $x_1 = 0$ and $x_2 = 10$ and $x_3 = 0$, $t_1 = 2$, $t_2 = 11$, $t_3 = 1$

when $x_1 = 0$ and $x_2 = 0$ and $x_3 = 10$, $t_1 = 2$, $t_2 = 1$, $t_3 = 6$

so all routes might be used.

$t_1 = t_2 = t_3$

$q = x_1 + x_2 + x_3 = 10$, so $x_2 = 10 - x_1 - x_3$

$t_1 = t_2$

$2 + 0.5x_1 = 1 + x_2 = 11 - x_1 - x_3$ or $2 + 1.5x_1 = 11 - x_3$

also $t_1 = t_3$

$2 + 0.5x_1 = 4 + 0.2x_3$ or $x_3 = (-2 + 0.5x_1)/0.2$

$x_3 = -10 + 2.5x_1$

So, $2 + 1.5x_1 = 11 - (-10 + 2.5x_1) = 21 - 2.5x_1$

$x_1 = 19/4 = 4.75$

and, $t_1 = 2 + 0.5(4.75) = 4.375$

$t_2 = 4.375 = 1 + x_2$

$x_2 = 3.375$

$t_3 = 4.375 = 4 + 0.2x_3$

$x_3 = 1.875$

Part b (with $q = 5000$):

when $x_1 = 5$ and $x_2 = 0$ and $x_3 = 0$, $t_1 = 4.5$, $t_2 = 1$, $t_3 = 4$

when $x_1 = 0$ and $x_2 = 5$ and $x_3 = 0$, $t_1 = 2$, $t_2 = 6$, $t_3 = 1$

when $x_1 = 0$ and $x_2 = 0$ and $x_3 = 5$, $t_1 = 2$, $t_2 = 1$, $t_3 = 5$

so all routes might be used.

routes 1 and 2 have the lowest free-flow travel times, so assume route 3 is not used.

$t_1 = t_2$ and $x_2 = 5 - x_1$

$2 + 0.5x_1 = 1 + x_2 = 1 + 5 - x_1$

$x_1 = 2.67$

$x_2 = 5 - 2.67 = 2.33$

Check to see if route 3 is used:

$t_1 = 2 + 0.5(2.67) = 3.335 < 4$ min (route 3 free-flow time). Therefore the assumption that route 3 is not used is valid.

PROBLEM 8.23

Flow on route 3 will begin when $t_1 = t_2 = 4$ (which is route 3's free-flow travel time).

so $t_1 = 2 + 0.5x_1 = 4$

$x_1 = 4$

$t_2 = 1 + x_2 = 4$

$x_2 = 3$

thus $q = 7$. If $q > 7$, flow on route 3. If $q \le 7$ flow on routes 1 and 2 only.

PROBLEM 8.24

The 2000 single occupant vehicles must use the unrestricted lanes. Total person hours can be written as (subscripting: r = restricted, u2 = 2 person using unrestricted, u1 = 1 person using unrestricted)

$z(x) = x_r t_r \times 2 + x_{u2} t_u \times 2 + x_{u1} t_u \times 1$

where $x_u = x_{u2} + x_{u1}$ and $x_{u1} = 2.0$

rewriting

$z(x) = 2x_r(4 + 2x_r) + 2x_{u2}(4 + 0.5(2.0 + x_{u2})) + 2(4 + 0.5(2.0 + x_{u2}))$

and $x_r + x_{u2} = 2.0$

$= 8x_r + 4x_r^2 + 10x_{u2} + x_{u2}^2 + 10 + x_{u2}$

$= 8(2 - x_{u2}) + 4(2 - x_{u2})^2 + 10x_{u2} + x_{u2}^2 + 10 + x_{u2}$

$= 5x_{u2}^2 - 13x_{u2} + 42$

$dz/dx_{u2} = 10x_{u2} - 13 = 0$

$x_{u2} = 1.3$ so, $x_r = 2.0 - 1.3 = 0.7$

Total person hours:

$t_r = 4 + 2(0.7) = 5.4$

$t_u = 4 + 0.5(3.3) = 5.65$

$2[5.4(700)] + 2[5.65(1300)] + 2000(5.65) = 33550$ person-min or 559.167 person-h

At UE, with 2000 vehicles on the unrestricted lanes, t_u can be written as,

$t_u = 4 + 0.5(2.0 + x_{u2}) = 5 + 5x_{u2}$

$x_{u2} + x_r = 2.0$

check if both lane choices are used:

when $x_u = 2$ and $x_r = 0$, $t_u = 7$, $t_r = 4$

when $x_u = 0$ and $x_r = 2$, $t_u = 5$, $t_r = 8$

so both lane choices might be used by 2 occupant vehicles

$t_u = t_r$

$5 + 0.5x_{u2} = 4 - 2x_r$

$x_r = 2 - x_{u2}$

$5 + 0.5x_{u2} = 4 - 2(2 - x_{u2})$

$x_{u2} = 1.2$

$x_r = 2 - 1.2 = 0.8$

$t_u = t_r = 5.6$

Total person hours:

$2[5.6(2000)] + 5.6(2000) = 33600$ person-min or 560 person-h. Savings in person hours is 0.833.

PROBLEM 8.25

a) open to all:

number of vehicles $= 2500 + 500 + 300 + 100 + 20 = 3420$

number of people $= 2500 + 500(2) + 300(3) + 100(4) + 20(50) = 5800$

$t = 15[1 + 1.15(3.42/3.6)^{6.87}] = 27.1269$

person-hours $= 27.1269(5800)/60 = 2622.27$

b) 2 + lane

SOVs

$t = 15[1 + 1.15(2.5/2.4)^{6.87}] = 37.83$

HOVs

$t = 15[1 + 1.15(0.92/1.2)^{6.87}] = 17.78$

person-hours = $[37.83(2500) + 17.78(3300)] = 2554.15$ person-h

c) 3 + lane

LOVs

$t = 15[1 + 1.15(3.0/2.4)^{6.87}]/60 = 94.90$ person-h

HOVs

$t = 15[1 + 1.15(0.42/1.2)^{6.87}] = 15.013$

person-hours = $[94.9(3500) + 15.013(2300)]/60 = 6111$ person-h

PROBLEM 8.26

a) 2 + lane

SOVs

$t = 15[1 + 1.15(2.0/2.4)^{6.87}] = 19.93$

HOVs

$t = 15[1 + 1.15(0.93/1.2)^{6.87}] = 17.99$

person-hours = $[19.93(2000) + 17.99(3800)] = 1803.97$ person-h

b) need 157,336 person-min. Let x be the number of shifters

$$\left[15 + 17.25\left(\frac{2.5 - x/1000}{2.4}\right)^{6.87}\right](2500 - x) + \left[15 + 17.25\left(\frac{0.92 + x/50000}{1.2}\right)^{6.87}\right](3300 + x) = 157336$$

$x = -22.17$

PROBLEM 8.27

At SO:

min $z(x) = (5 + 3x_1)x_1 + (7 + 3x_2)x_2$

$x_2 = 7 - x_1$

$z(x) = (5 + 3x_1)x_1 + (7 + 3(7 - x_1))(7 - x_1)$

$dz/dx_1 = 5 + 6x_1 - 7 - 14 + 2x_1 = 0$

$x_1 = 2$ and $x_2 = 5$

travel time is $11(2000) + 12(5000) = 82000$ veh-min

at UE:

$t_1 = t_2$

$5 + 3x_1 = 7 + (7 - x_1)$

$x_1 = 2.25$

$x_2 = 7 - 2.25 = 4.75$

travel time is $11.75(7000) = 82250$ veh-min

PROBLEM 8.28

$z(x) = \int(5 + 3w)\, dw + \int(7 + 3w)\, dw$

and $x_2 = 7 - x_1$

$z(x) = 5x_1 + 1.5x_1^2 + 49 - 7x_1 + 24.5 - 7x_1 + 0.5x_1^2$

$dz/dx_1 = 4x_1 - 9$

at $x_1 = 2$ (the SO solution)

$dz/dx_1 = -1$

PROBLEM 8.29

when $x_1 = 6$ and $x_2 = 0$, $t_1 = 16.5$ and $t_2 = 5$

when $x_1 = 0$ and $x_2 = 6$, $t_1 = 3$ and $t_2 = 21$

so both routes might be used

Existing conditions with, $t_1 = t_2$ and $x_2 = 6 - x_1$

81

$3 + 1.5(x_1/2)^2 = 5 + 2.67(6 - x_1)$

$0.375x_1^2 + 2.67x_1 - 18.02 = 0$

$x_1 = 4.233$

$t_1 = 3 + 1.5(4.233/2)^2 = 9.719 = t_2$

After capacity expansion, $t_2 = 5 + 1.6x_2$

and $q = 6 + 0.5(9.719 - (5 + 1.6x_2)) = 8.3595 - 0.8x_2$

$q = x_1 + x_2 = 8.3595 - 0.8x_2$

$x_2 = (8.3595 - x_1)/1.8 = 4.644 - 0.556x_1$

$t_1 = t_2$

$3 + 0.375x_1^2 = 5 + 1.6(4.644 - 0.556x_1)$

$0.375x_1^2 + 0.889x_1 - 9.43 = 0$

$x_1 = 3.968$

$t_1 = 3 + 1.5(3.968/2)^2 = 8.903 = t_2$

$x_2 = (8.3595 - 3.968)/1.8 = 2.4395$

$q = 6.406 (2.4395 + 3.968)$

PROBLEM 8.30

<u>At SO:</u>

$\min z(x) = 5x_1 + 4x_1^2 + 7x_2 + 2x_2^2$

$x_2 = q - x_1$

$z(x) = 5x_1 + 4x_1^2 + 7(q - x_1) + 2(q - x_1)^2$

$dz/dx_1 = -7 - 4q + 4x_1 + 5 + 8x_1 = 0$

$x_1 = 0.3333q + 0.667$

<u>at UE:</u>

$t_1 = t_2$

$5 + 4x_1 = 7 + 2(q - x_1)$

$6x_1 - 2q - 2 = 0$

substituting,

$1.3333q + 1.3333 - 2q - 2 = 0$

$q = -1$, so is not possible.

PROBLEM 8.31

when $x_1 = 4$ and $x_2 = 0$ and $x_3 = 0$, $t_1 = 11$, $t_2 = 12$, $t_3 = 2$

when $x_1 = 0$ and $x_2 = 4$ and $x_3 = 0$, $t_1 = 5$, $t_2 = 24$, $t_3 = 12$

when $x_1 = 0$ and $x_2 = 0$ and $x_3 = 4$, $t_1 = 5$, $t_2 = 12$, $t_3 = 5.2$

so only routes 1 and 3 will be used

$t_1 = t_3$ and $x_1 = 4 - x_3$

$5 + 1.5(4 - x_3) = 2 + 0.2x_3^2$

$0.2x_3^2 + 1.5x_3 - 9 = 0$

$x_3 = 3.935$, $x_1 = 4 - 3.935 = 0.065$

PROBLEM 8.32

$t_1 = 6 + 4x_1$

$t_2 = 2 + 0.5x_2^2$

$q - 0.1(t_2 - 2) = x_1 + x_2$

$q - 0.1((2 + 0.5x_2^2) - 2) = x_1 + x_2$

$x_1 = 4 - x_2 - 0.05x_2^2$

$t_1 = 6 + 4x_1 = t_2 = 2 + 0.5x_2^2$

$2 + 0.5x_2^2 = 6 + 4(4 - x_2 - 0.05x_2^2)$

$0.7x_2^2 + 4x_2 - 20 = 0$

$x_2 = 3.203$

$t_2 = 2 + 0.5(3.203)^2 = 7.132$

$q - 0.1(7.132 - 2) = 3.487$

$x_1 = 3.487 - 3.203 = 0.284$

Total travel time $= 3.487(7.132) = 24.870$ or 24,870 vehicle-min

PROBLEM 8.33

(subscripting: r = restricted, u2 = 2 person using unrestricted, u1 = 1 person using unrestricted)

$z(x) = x_r t_r \times 2 + x_{u2} t_u \times 2 + x_{u1} t_u \times 1$

where $x_u = x_{u2} + x_{u1}$ and $x_{u1} = 2.0$

rewriting

$z(x) = 2x_r(12 + x_r) + 2x_{u2}(12 + 0.5(3.0 + x_{u2})) + 3(12 + 0.5(3.0 + x_{u2}))$

and $x_r + x_{u2} = 4.0$

$= 24x_r + 2x_r^2 + 24x_{u2} + 3x_{u2} + x_{u2}^2 + 36 + 4.5 + 1.5x_{u2}$

$= 3x_{u2}^2 - 11.5x_{u2} + 168.5$

$dz/dx_{u2} = 6x_{u2} - 11.5 = 0$

$x_{u2} = 1.916$ so, $x_r = 4.0 - 1.916 = 2.083$, $x_u = 4.916$

Total person hours:

$t_r = 12 + 2.083 = 14.083$

$t_u = 12 + 0.5(4.916) = 14.458$

$2[2.083(14.083)] + 2[1.916(14.458)] + 3(14.458) = 157.444$ or $157,444$ person-min

PROBLEM 8.34

at UE:

$t_1 = t_2$ and $x_1 = 3x_2$

$5 + (3x_2/2)^2 = 7 + (x_2/4)^2$

$x_2 = 0.956$, $x_1 = 2.869$, $q = 3.825$

Total travel time = $[5 + (3(0.914)/2)^2] \times 3825/60 = 449.93$ veh-h

At SO:

$\min z(x) = (5 + (x_1/2)^2)x_1 + (7 + (x_2/4)^2)x_2$

$x_1 = 3.825 - x_2$

$z(x) = (5 + ((3.825 - x_2)/2)^2)(3.825 - x_2) + (7 + (x_2/4)^2)x_2$

$z(x) = -0.187x_2^3 + 2.869x_2^2 - 8.9752x_2 + 33.117$

$dz/dx_2 = -0.561x_2{}^2 + 5.738x_2 - 8.975 = 0$

$x_2 = 1.927$

$x_1 = 3.825 - 1.927 = 1.898$

Total travel time:

$t_1 = 5 + (1.898/2)^2 = 5.901$ min

$t_2 = 7 + (1.927/2)^2 = 7.232$ min

Total travel time = $[(1898(5.901)) + (1927(7.232))]/60 = 418.94$ veh-h

Savings = $449.93 - 418.94 = 30.99$ veh-h

PROBLEM 8.35

at UE:

with $x_1 = q - x_2$

$$= \int_0^{q-x_2} (5 + 3.5w)dw + \int_0^{x_2} (1 + 0.5w^2)dw$$

$$= (5w + 1.75w^2)\big|_0^{q-x_2} + (w + 0.167w^3)\big|_0^{x_2}$$

$$= 5q - 5x_2 + 1.75q^2 - 3.5qx_2 + 1.75x_2^2 + x_2 + 0.167x_2^3$$

$$\frac{dz}{dx_2} = -5 - 3.5q + 3.5x_2 + 1 + 0.5x_2^2 = 0.5x_2^2 + 3.5x_2 - 4 - 3.5q$$

At SO:

min $z(x) = (5x_1 + 3.5x_1{}^2) + (x_2 + 0.5x_2{}^2)$

or with $x_1 = q - x_2$

$z(x) = 5q - 5x_2 + 3.5q^2 - 7qx_2 + 3.5x^2 + x_2 + 0.5x_2{}^3$

$dz/dx_2 = 1.5x_2{}^2 + 7x_2 - 4 - 7q = 0$

SO dz/dx_2 - UE $dz/dx_2 = 7$

$(1.5x_2{}^2 + 7x_2 - 4 - 7q) - (0.5x_2^2 + 3.5x_2 - 4 - 3.5q) = 7$

with $x_2 = 3$

$q = 3.57$

<u>UE:</u>

$$0.5x_2^2 + 3.5x_2 - 4 - 3.5(3.57) = 0$$

gives $x_2 = 3.226$ (by quadratic) so $x_1 = 3.57 - 3.226 = 0.344$

Total travel time $= [5 + 3.5(0.344)](3.57)(1000) = 22,134$ veh-min

<u>SO:</u>

$$1.5x_2^2 + 7x_2 - 4 - 7(3.57) = 0$$

gives $x_2 = 2.64$ (by quadratic) so $x_1 = 3.57 - 2.64 = 0.927$

$tt_1 = [5 + 3.5(0.927)]0.927 \times 1000 = 7,642$ veh-min

$tt_2 = [1 + 0.5(2.64)^2]2.64 \times 1000 = 11,840$ veh-min

or 19,480 veh-min total

SOLUTIONS: APPENDIX A

PROBLEM A.1

$$a = \frac{G_2 - G_1}{2L} = \frac{6.5 + 3.5}{2(16)} = 0.3125$$

$$b = -3.5$$

at low point

$$\frac{dy}{dx} = 2ax + b = 0.625x - 3.5 = 0$$

$$x = 5.6 \text{ sta}$$

for elevation of low point

$$y = ax^2 + bx + c = 0.3125(5.6)^2 - 3.5(5.6) + 1500 = 1490.2 \text{ ft}$$

PVI sta $120+00 + 8+00 = 128+00$

   elev $1500 - 3.5(8) = 1472$ ft

PVT sta $120+00 + 16+00 = 136+00$

   elev $1472 + 6.5(8) = 1524$ ft

low point sta $120+00 + 5+60 = 125+60$; elev 1490.2 ft

PROBLEM A.2

$$a = \frac{G_2 - G_1}{2L} = \frac{-2.5 - 4}{2(5)} = -0.65$$

$$b = 4$$

at high point

$$\frac{dy}{dx} = 2ax + b = (2)(-0.65)x + 4 = 0$$

$$x = 3.077 \text{ sta}$$

for elevation of high point

$$c = 1322 - 4(2.5) = 1312 \text{ ft}$$

$$y = ax^2 + bx + c = -6.15 + 12.31 + 1312 = 1318.16 \text{ ft}$$

PVC sta $340+00 - 2+50 = 337+50$

    elev 1312 ft

PVT sta $340+00 + 2+50 = 342+50$

    elev $1322 - 2.5(2.5) = 1315.75$ ft

high point sta $337+50 + 3+07.7 = 340+57.7$

       elev 1318.16 ft

PROBLEM A.3

$$a = \frac{G_2 - G_1}{2L} = \frac{-1.08 - 1.2}{2(6)} = -0.19$$

$b = 1.2$

PVC sta $110+00 - 3+00 = 107+00$

    elev $1098.4 - 3(1.2) = 1094.8$ ft

at station $110+85$ the pipe is 385 ft from the PVC $(110+85 - 107+00)$

curve elevation

$y = ax^2 + bx + c$

$y = -0.19(3.85)^2 + 1.2(3.85) + 1094.8$

$y = 1096.6$ ft

since the top of the pipe is at 1093.6 ft, the pipe is

3 ft below the surface

PROBLEM A.4

for crest curve, K at 60 mi / h is 310 desirable

$A = |G_2 - G_1| = 2.28$

$L_m = 310(2.28) = 706.8$ ft so distance is not adequate

PROBLEM A.5

$$L = 2(110+92.5 - 109+00) = 385 \text{ ft}$$

$$G_1 = \frac{\text{elev PVC - elev PVI}}{L/2} = -\frac{950 - 947.11}{192.5} = -0.015$$

$$x = K|G_1|$$

$$K = \frac{x}{|G_1|} = \frac{110+65 - 109+00}{|1.5|} = 110$$

from Table A.3, the design speed is 50 mi / h for $K = 110$

PROBLEM A.6

$$A = |G_2 - G_1| = 3.0$$

at 70 mi / h, $K = 540$ for crest desirable (Table A.2)

$$L = KA = 540(3) = 1620 \text{ ft}$$

f goes from 0.28 to 0.392 (1.4×0.28), and reaction time goes from 2.5 to 3 s (2.5×1.2). Using Eq. A.12

$$\text{SSD} = \frac{V_1^2}{2g(f \pm G)} + V_1 t_r = \frac{(70 \times 1.47)^2}{2(32.2)(0.392)} + (70 \times 1.47)3.0 = 728.13 \text{ ft}$$

assuming $L > S$ Eq A.14 gives

$$L_m = \frac{AS^2}{200\left(\sqrt{H_1} + \sqrt{H_2}\right)^2} = \frac{3(728.13)^2}{200\left(\sqrt{2.75} + \sqrt{0.25}\right)^2} = 1707.7 \text{ ft}$$

the 2030 curve is 87.7 ft longer

PROBLEM A.7

$$Y = \frac{A}{200L} x^2$$

$$A = \frac{(3)(200)(800)}{(352)^2} = 3.87$$

$$K \text{ existing} = \frac{L}{A} = \frac{800}{3.87} = 206.5$$

which exceeds the minimum (190). At desirable (310) as shown in Table A.2 acceptable length for desirable is $L = (310)(3.87) = 1199.7$ ft

PROBLEM A.8

$K_c = 310$ (Table A.2) 60 mi / h desirable

$K_s = 160$ (Table A.3) 60 mi / h desirable

$A = |0 - 2| = 2$

$L_c = 310(2) = 620$ ft

$L_s = 160(2) = 320$ ft

for crest curve,

sta $PVT_c = 0+00 + 6+20 = 6+20$

elev $PVT_c = 100 - \dfrac{AL_c}{200} = 100 - \dfrac{(2)(620)}{200} = 93.8$ ft

for sag curve,

sta $PVT_s = \dfrac{L_c}{2} + 40+00 + \dfrac{L_s}{2} = 44+70$

elev $PVT_s = 100 - 0.02(4000) = 20$ ft

sta $PVC_s = 44+70 - 3+20 = 41+50$

elev $PVC_s = 20 + \dfrac{AL_s}{200} = 20 + \dfrac{(2)(320)}{200} = 23.2$ ft

PROBLEM A.9

$K_c = 310$ (Table A.2) 60 mi / h desirable

$K_s = 160$ (Table A.3) 60 mi / h desirable

drop in elevation is $0.02(4000) = 80$ ft

$$80 = \frac{AL_c}{200} + \frac{AL_s}{200} + \frac{A(4000 - L_c - L_s)}{100}$$

substituting $K's$

$$80 = \frac{310A^2}{200} + \frac{160A^2}{200} + 40A - 3.1A^2 - 1.6A^2$$

$A = 2.315$ (the smaller of the roots)

$PVC_c = $ sta $0+00$; elev 100 ft

PVT_c sta $= 0+000 + 310(2.315) = 7+17.65$

$$PVT_c \text{ elev} = 100 - \frac{AL_c}{200} = 100 - 8.306 = 91.694 \text{ ft}$$

$$PVC_s \text{ elev} = 91.694 - \frac{A(4000 - L_c - L_s)}{100} = 91.694 - 67.411 = 24.283 \text{ ft}$$

PVC_s sta $= 7+17.65 - (4000 - L_c - L_s) = 36+29.6$

$$PVT_s \text{ elev} = 24.283 - \frac{AL_s}{200} = 20 \text{ ft}$$

PVT_s sta $= 40+00$

PROBLEM A.10

From Table A.2, $K = 310$. The high point is

$x = K|G_1| = 310(4) = 1240$ ft

at the high point

$$\frac{dy}{dx} = 2ax + b = 2a(12.4) + 4 = 0$$

$a = -0.1613$

$y = ax^2 + bx + c$, with $c = 0$

$y = -0.1613(12.4)^2 + 0.04(12.4) = 24.8$ ft

PROBLEM A.11

$A = 1.0 + 0.5 = 1.5$

$L = 2(5744 - 5484) = 520$ ft

$K = \dfrac{L}{A} = \dfrac{520}{1.5} = 346.67$

At 55 mi / h we need $K = 1230$ (Table A.4), since $346.93 < 1230$ the curve is not long enough. The curve would have to be

$L = KA = 1230(1.5) = 1845$ ft

PROBLEM A.12

$24 = \dfrac{AL_c}{200} + \dfrac{AL_s}{200}$

$K_c = 160;\ K_s = 110$

$\dfrac{160A^2}{200} + \dfrac{110A^2}{200} = 24$

$A = 4.216\%;\ L_c = 160(4.216) = 674.6$ ft; $L_s = 110(4.216) = 463.776$ ft

so total distance $= 2{,}476.72$ ft

PROBLEM A.13

at 40 mi / h, $K_c = 80$, $K_s = 70$; and $L = 3500$ ft

Δelev $= 41$ ft

$Y_{fc} + Y_{fs} + \Delta x = \Delta Elev + \Delta G_c$

$\dfrac{(4+x)^2 80}{200} + \dfrac{x^2 70}{200} + \dfrac{x\big(3500 - (4+x)80 - 70x\big)}{100} = 41 + 0.04\big((4+x)80\big)$

$x = 1.547\%$

$L_c = 18(5.547) = 443.76$ ft

$L_s = 18(1.547) = 108.29$ ft

$L_{constant} = 2947.95$ ft

PROBLEM A.14

$K_c = 160$

$A = \dfrac{L}{K_c} = \dfrac{1200}{60} = 7.5$

S at 60 mi / h is 650 ft

with $L > S$ (Eq A.14)

$L_m = \dfrac{AS^2}{200\left(\sqrt{H_1} + \sqrt{H_2}\right)^2} = 1200 = \dfrac{7.5(650)^2}{200\left(\sqrt{H_1} + 0.707\right)^2}$

$H_1 = 8.56$ ft

PROBLEM A.15

at 50 mi / h, $K_c = 160$, $K_s = 110$;

$L_c = 8(160) = 1280$ ft

$L_s = 8(110) = 770$ ft

$Y_{fc} = \dfrac{A_c L}{200} = \dfrac{8(1280)}{200} = 51.2$ ft

$Y_{fs} = \dfrac{A_s L}{200} = \dfrac{7(770)}{200} = 26.95$ ft

$\Delta x = (3000 - 2050)(0.05) = 47.5$ ft

$\Delta G_c = 0.03(1280) = 38.4$ ft

$\Delta G_s = 0.02(770) = 15.4$ ft

$\Delta \text{Elev} = Y_{fc} + Y_{fs} + \Delta x - \Delta G_c - \Delta G_s$

$\Delta \text{Elev} = 51.2 + 26.95 + 47.5 - 38.4 - 15.4$

$\Delta \text{Elev} = 71.85$ ft

PROBLEM A.16

at 40 mi / h SSD $=$ 325 ft

at 50 mi / h SSD $=$ 475 ft

$M_s = 34 - 6 = 28$ ft

try 40 mi / h; $f_s = 0.15$

$$R_v = \frac{V^2}{g(f_s + e)} = \frac{(40 \times 1.47)^2}{32.2(0.23)} = 466.8 \text{ ft}$$

$$M_s = R_v\left(1 - \cos\frac{90\,SSD}{\pi R_v}\right) = 466.8\left(1 - \cos\frac{90\,(325)}{\pi(466.8)}\right) = 28 \text{ ft} \therefore \text{ good}$$

try 50 mi / h; $f_s = 0.14$

$$R_v = \frac{V^2}{g(f_s + e)} = \frac{(50 \times 1.47)^2}{32.2(0.22)} = 762.6 \text{ ft}$$

$$M_s = R_v\left(1 - \cos\frac{90\,SSD}{\pi R_v}\right) = 762.6\left(1 - \cos\frac{90\,(475)}{\pi(762.6)}\right) = 36.69 \text{ ft} \therefore \text{ no good}$$

use a 40 mi / h design speed

PROBLEM A.17

$$R_v = \frac{V^2}{g(f_s + e)} = \frac{(70 \times 1.47)^2}{32.2(0.16)} = 2055.2 \text{ ft}$$

$T = 131+40 - 124+10 = 730$ ft

$$T = R \tan \frac{\Delta}{2} = 730 = 2055.2 \tan \frac{\Delta}{2}$$

$\Delta = 39.09$

$$L = \frac{100(32.82)}{\dfrac{5729.6}{2055.2}} = 1402.15 \text{ ft}$$

PT sta $124+10 + 14+02.15 = 138+12.15$

94

PROBLEM A.18

$$\text{sta } PC = 2700 + 00 - 5 + 10 = 2694 + 90$$

$$R = \frac{T}{\tan \dfrac{\Delta}{2}} = \frac{510}{\tan 20} = 1401.2 \text{ ft}$$

$$D = \frac{18000}{\pi R} = \frac{5729.6}{1401.2} = 4.089$$

$$L = \frac{100\Delta}{D} = \frac{100(40)}{4.089} = 978.23 \text{ ft}$$

$$\text{sta } PT = 2694 + 90 + 9 + 78.23 = 2704 + 68.23$$

for design speed R_v is $1401.2 - 15 = 1386.2$ ft

$$= \sqrt{gR_v(f_s + e)} = \sqrt{32.2(1386.2)(0.17)} = 87.1 \text{ ft / s or } 59.26 \text{ mi / h}$$

PROBLEM A.19

$$e = \frac{V^2}{gR_v} - f_s = \frac{(70 \times 1.47)^2}{32.2(900)} - 0.1 = 0.265 \text{ which exceeds allowable maximum}$$

PROBLEM A.20

$$\tan \alpha = \frac{\dfrac{V^2}{gR_v} - f_s}{1 + f_s \dfrac{V^2}{gR_v}} = \frac{\dfrac{(100 \times 1.47)^2}{32.2(1000)} - 0.20}{1 + 0.20\left(\dfrac{(100 \times 1.47)^2}{32.2(1000)}\right)} = 0.415 \text{ ft / ft}$$

$$D = \frac{18000}{\pi R} = \frac{5729.6}{1000} = 5.7296$$

$$L = \frac{100\Delta}{D} = \frac{100(30)}{5.7296} = 523.6 \text{ ft}$$

$$T = R \tan \frac{\Delta}{2} = 1000 \tan \frac{30}{2} = 267.95 \text{ ft}$$

$$\text{sta } PC = 1125 + 10 - 2 + 67.95 = 1122 + 42.05$$

$$\text{sta } PT = 1122 + 42.05 - 5 + 23.60 = 1127 + 65.65$$

PROBLEM A.21

at 65 mi / h; $f_s = 0.11$

$$R_v = \frac{V^2}{g(f_s + e)} = \frac{(65 \times 1.47)^2}{32.2(0.19)} = 1492.28 \text{ ft}$$

radius to centerline is $1492.28 + 6 = 1498.28$ ft

$$D = \frac{18000}{\pi(1498.28)} = 3.82$$

$$T = R \tan \frac{\Delta}{2} = (1498.28) \tan 17.5 = 472.41 \text{ ft}$$

$$L = \frac{100(35)}{3.82} = 916.23 \text{ ft}$$

PC sta $250 + 50 - 4 + 72.41 = 245 + 77.59$

PT sta $245 + 77.59 + 9 + 16.23 = 254 + 93.82$

PROBLEM A.22

at 70 mi / h; $f_s = 0.1$

$$R_v = \frac{V^2}{g(f_s + e)} = \frac{(70 \times 1.47)^2}{32.2(0.16)} = 2055.20 \text{ ft}$$

radius to centerline is $2055.20 + 5 = 2060.20$ ft

$$D = \frac{18000}{\pi(2060.20)} = 2.781$$

$$L = \frac{100(40)}{2.781} = 1438.33 \text{ ft}$$

PROBLEM A.23

from Eq A.35 $R_v = R = \dfrac{180L}{\pi\Delta} = \dfrac{180(628)}{\pi 90} = 399.8$ ft

from Eq A.40 with $M_s = 19.4$ ft

$$SSD = \frac{\pi R_v}{90}\left[\cos^{-1}\left(\frac{R_v - M_s}{R_v}\right)\right] = \frac{\pi(399.8)}{90}\left[\cos^{-1}\left(\frac{399.8 - 19.4}{399.8}\right)\right] = 250.1\ \text{ft}$$

looking at Table A.1, the design speed is about 35 mi / h desirable

PROBLEM A.24

from Problem A.21 $R_v = 1492.28$ ft

from Table A.1 at 65 mi / h

SSD = 550 ft (minimum)

SSD = 725 ft (desirable)

for mimimum using Eq A.39

$$M_s = R_v\left(1 - \cos\frac{90\ SSD}{\pi R_v}\right) = 1492.28\left(1 - \cos\frac{90\ (550)}{\pi(1492.28)}\right) = 25.27\ \text{ft}$$

or 25.27 - 6 = 19.27 ft from the inside lane

for desirable using Eq A.39

$$M_s = R_v\left(1 - \cos\frac{90\ SSD}{\pi R_v}\right) = 1492.28\left(1 - \cos\frac{90\ (725)}{\pi(1492.28)}\right) = 43.82\ \text{ft}$$

or 43.82 - 6 = 37.82 ft from the inside lane